For Bowie —C. S.

For my family and friends,
for accompanying me during this trip —X. A.

Text copyright © 2018 by Colin Stuart
Illustrations copyright © 2018 by Ximo Abadía
Design copyright © 2018 by Kings Road Publishing Limited

First U.S. edition 2019
First published in the U.K. in 2018 by Big Picture Press

Library of Congress Catalog Card Number pending
ISBN 978-1-5362-0855-9

19 20 21 22 23 24 TWP 10 9 8 7 6 5 4 3 2 1

Printed in Johor Bahru, Malaysia

This book was typeset in Ulissa and BsKombat.
The illustrations were created with graphite,
wax, and ink and colored digitally.

BIG PICTURE PRESS
an imprint of
Candlewick Press
99 Dover Street
Somerville, Massachusetts 02144

www.candlewick.com

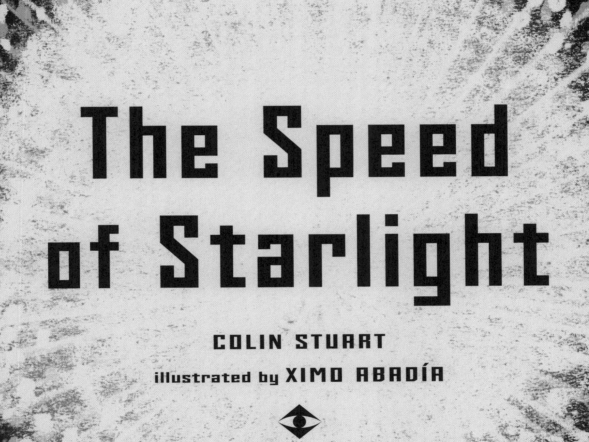

The Speed of Starlight

COLIN STUART

illustrated by XIMO ABADÍA

BPP

PHYSICS

SOUND

LIGHT AND COLOR

SPACE

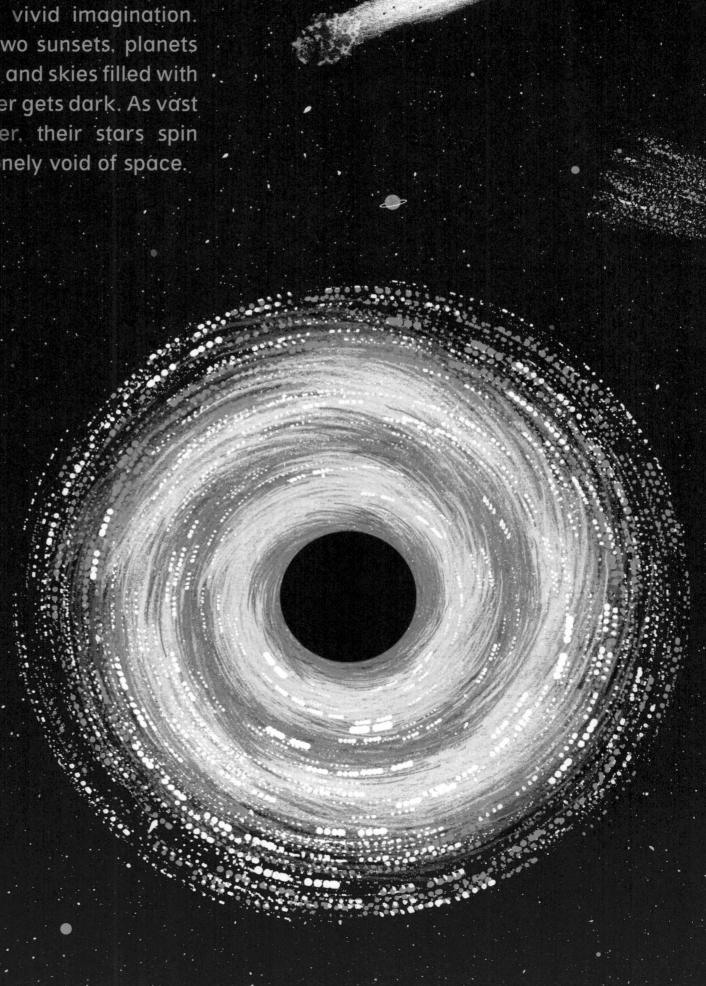

Welcome to the Universe

Our universe is a truly remarkable place, rivaling even the most vivid imagination. There are worlds with two sunsets, planets where it rains diamonds, and skies filled with so many stars that it never gets dark. As vast galaxies smash together, their stars spin dizzyingly out into the lonely void of space.

BLACK HOLES twist and warp not only space but time, too. **STARS** explode with such unimaginable force that they can outshine the light of a billion of their neighbors combined.

A dying star can become so squeezed that a single teaspoon of it weighs more than every person on Earth put together.

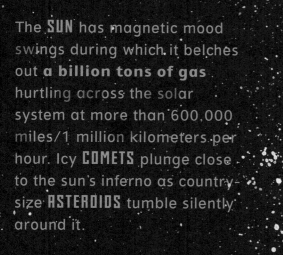

The **SUN** has magnetic mood swings during which it belches out **a billion tons of gas** hurtling across the solar system at more than 600,000 miles/1 million kilometers per hour. Icy **COMETS** plunge close to the sun's inferno as country-size **ASTEROIDS** tumble silently around it.

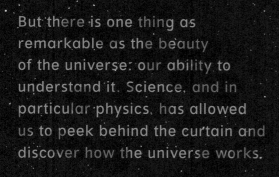

But there is one thing as remarkable as the beauty of the universe: our ability to understand it. Science, and in particular physics, has allowed us to peek behind the curtain and discover how the universe works.

What Is Physics?

Physics is the science of **ENERGY**, **MATTER**, and **FORCES**. Think of it like a cookbook for the cosmos. It lists all the ingredients — the forces and particles — needed to make everything around us. It even tells us how to combine them in different ways to explain why objects behave the way they do. Physics explains everything from an electric circuit to the birth of our universe.

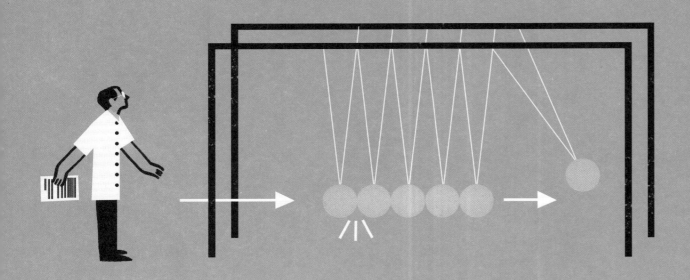

Physics can be divided into two main sets of rules: **QUANTUM PHYSICS**, which describes **very small** things and forces, and Einstein's **GENERAL THEORY OF RELATIVITY**, which describes **very big** things and forces. Physicists would love to combine these two theories into one "theory of everything," but so far that task has proven to be very difficult.

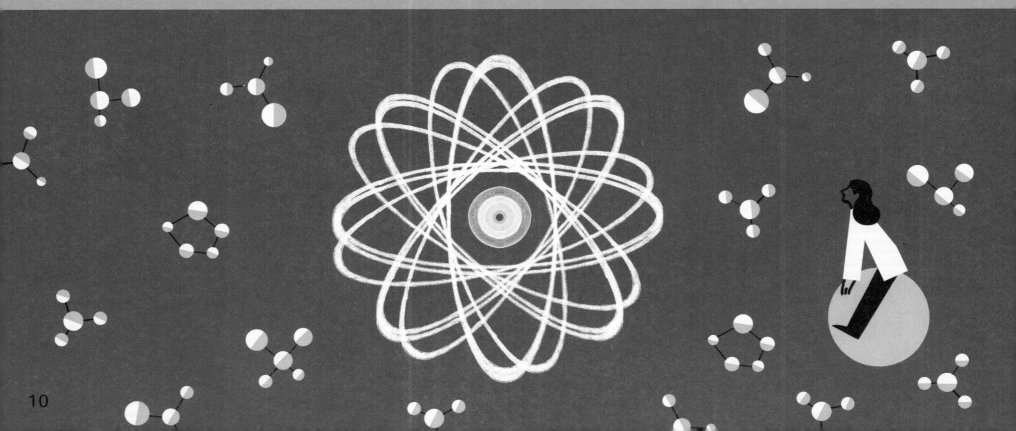

There are also some aspects of the universe that physicists don't currently know how to explain. For example, there seems to be an invisible glue called dark matter (page 71) holding galaxies together, but we don't know what it is made of. So physics is just as much about trying to solve new mysteries as it is about successfully explaining old ones. Physics is never finished.

Newton's Laws of Motion

A person who studies physics is called a **physicist**, and **ISAAC NEWTON** is one of the most famous physicists of all time. He did very important work on forces — the interactions that happen when objects move or have an effect on one another.

He came up with a theory about the force of gravity, and he also put together three rules for **how objects move**. They are called NEWTON'S THREE LAWS OF MOTION.

220 lb. / 100 kg

LAW ONE

Unless a force is acting on it, a stationary object will remain stationary and an object traveling at a constant speed will continue at that speed in the same direction.

LAW TWO

The more force applied to an object, the more it accelerates.

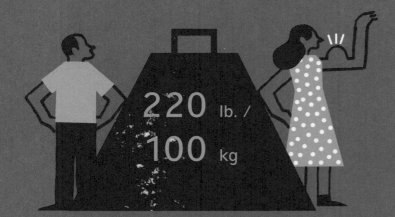

220 lb. / 100 kg

220 100

LAW THREE

For every action, there is an equal and opposite reaction.

These laws of motion help us send rockets into space. To get a stationary rocket moving, we need to **first apply a force [LAW ONE]. LAW TWO** tells us **how much force** we need to apply, and **LAW THREE** tells us **where we need to apply it**. So, for the rocket to take off, we need to fire something out of the bottom of it.

Down to Earth

High above the surface of the Earth, you jump out of a plane. As you hurtle toward the ground, the wind rushes past your ears and the sound is deafening. Then you open your parachute, and everything falls silent as you drift toward a safe landing.

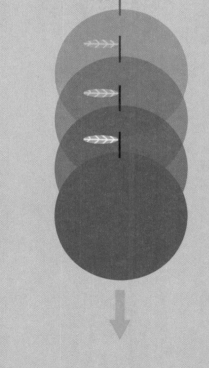

Sky divers fall to earth because the entire mass of the planet is pulling them down. Isaac Newton's genius was to realize that the moon orbits the Earth for the same reason an apple (or a skydiver) falls: an attractive force called **GRAVITY**. This is the **invisible force that pulls objects toward one another**.

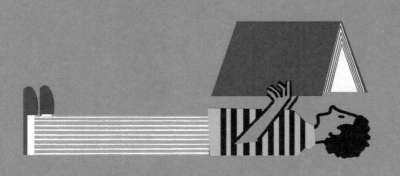

If any two objects have a gravitational attraction toward each other, then that means you are gravitationally attracted to this book. So why aren't you being pulled closer to it? Well, the strength of the gravitational force is not enough to overcome the force of FRICTION between the book and your hands. Friction is **the force that slows down objects that rub against each other**.

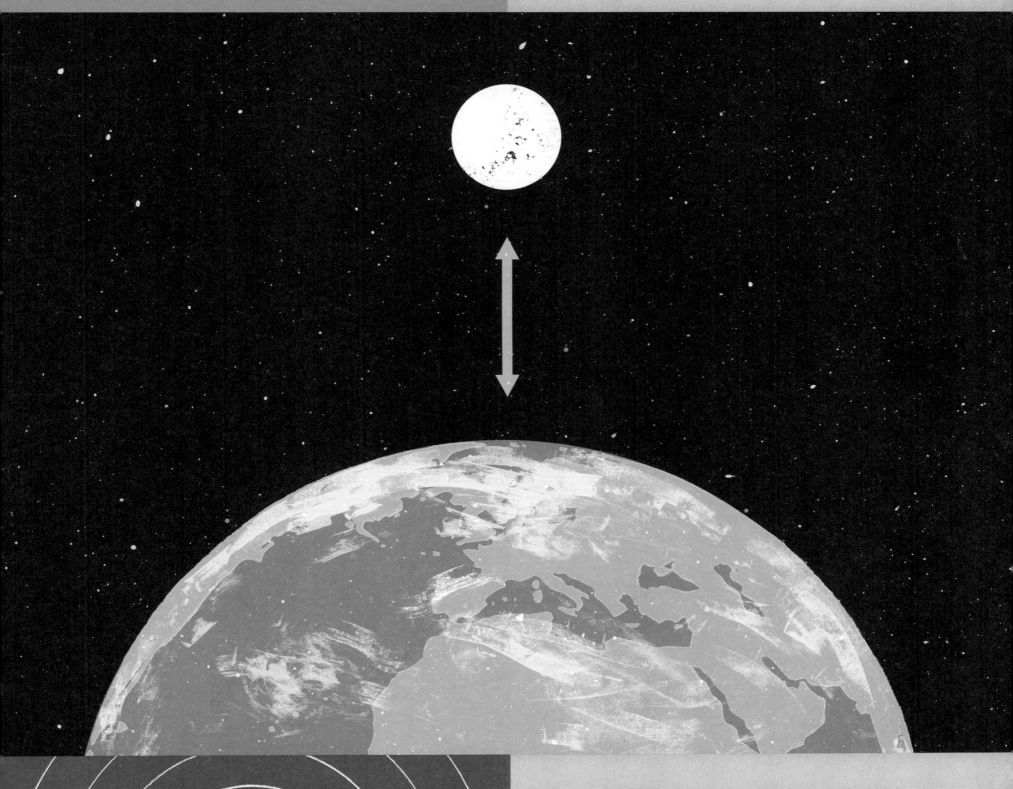

WHY DOESN'T THE MOON FALL DOWN?

Actually the moon is always falling, but it is moving at such a perfect speed that it continually circles the Earth. Any faster, and it would shoot off into space; any slower, and it would crash into us! Gravity is also what holds all the planets in our solar system in orbit around the sun.

Electricity and Magnetism

CRASH! BANG! BOOM! As a thunderstorm rages and rumbles overhead, suddenly a fork of lightning tears toward the ground and lights up the sky. This is nature at its fiercest and is an impressive display of the power of electricity.

ELECTRICITY is a type of energy that comes from the **flow of electric charge**. When **electric charges move**, they also create something else: **MAGNETISM**. Electricity and magnetism are so closely related that physicists describe them as one force, called the **ELECTROMAGNETIC FORCE**. It is stronger than gravity, which is why you can lift up one magnet with another.

electromagnetic force

gravity

Without magnetism, you wouldn't be here. The Earth has a giant **MAGNETIC FORCE FIELD** around it thanks to **electrical charges moving around inside our planet's molten core**. This shield deflects a lot of particles coming down from space that are dangerous to living things. It's one of the reasons sending people to Mars is difficult: we have to invent a way (such as protective pods) to shield astronauts from all those particles after they leave Earth's friendly magnetic field.

Earth's Force Field

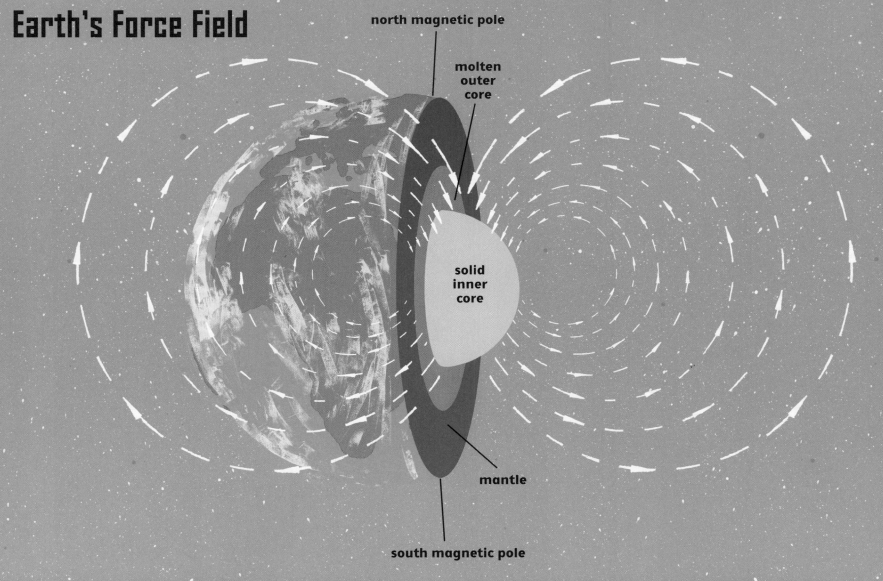

north magnetic pole

molten outer core

solid inner core

mantle

south magnetic pole

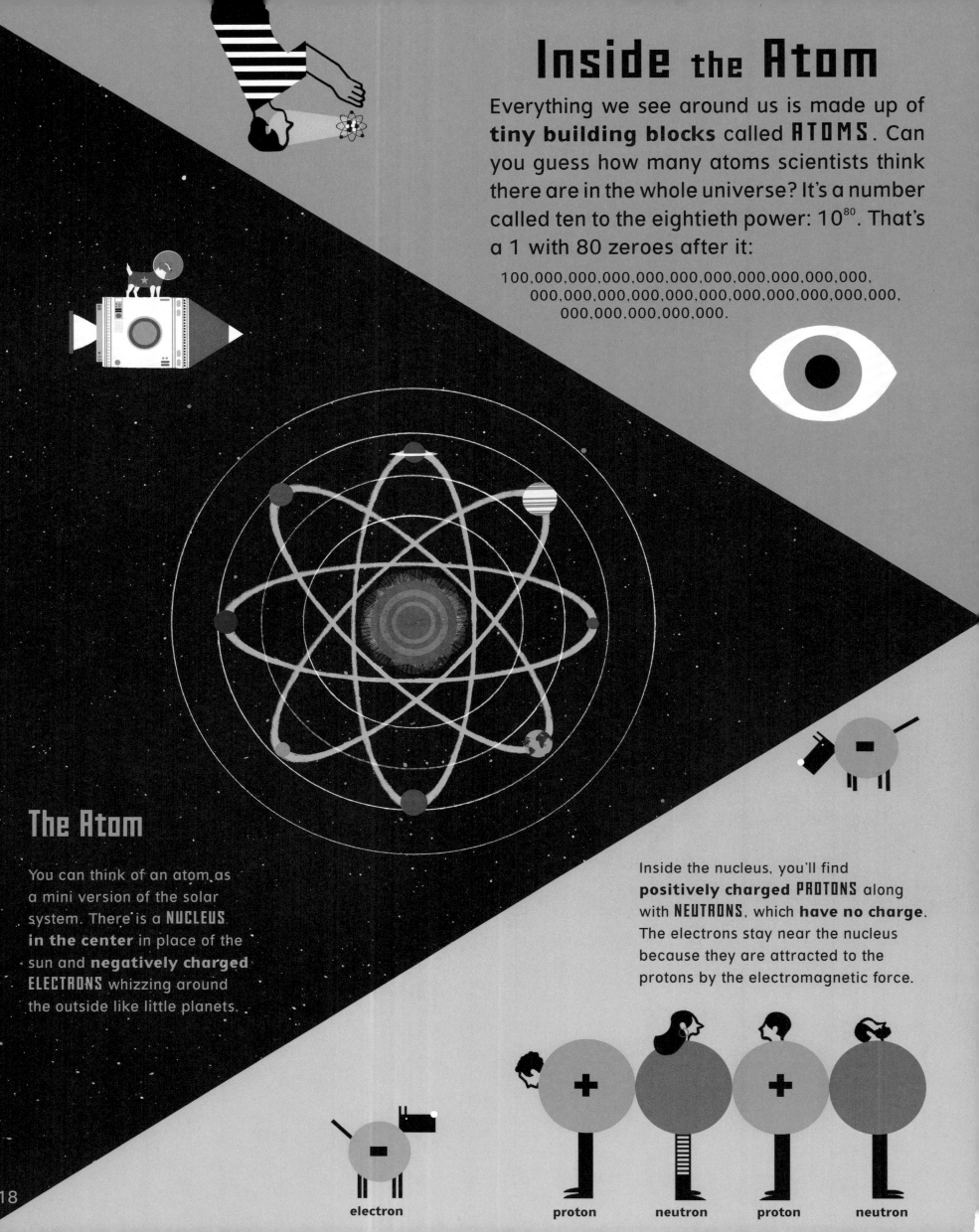

Inside the Atom

Everything we see around us is made up of **tiny building blocks** called **ATOMS**. Can you guess how many atoms scientists think there are in the whole universe? It's a number called ten to the eightieth power: 10^{80}. That's a 1 with 80 zeroes after it:

100,000,000,000,000,000,000,000,000,000,
000,000,000,000,000,000,000,000,000,000,000,
000,000,000,000,000.

The Atom

You can think of an atom as a mini version of the solar system. There is a **NUCLEUS** **in the center** in place of the sun and **negatively charged ELECTRONS** whizzing around the outside like little planets.

Inside the nucleus, you'll find **positively charged PROTONS** along with **NEUTRONS**, which **have no charge**. The electrons stay near the nucleus because they are attracted to the protons by the electromagnetic force.

electron

proton   neutron   proton   neutron

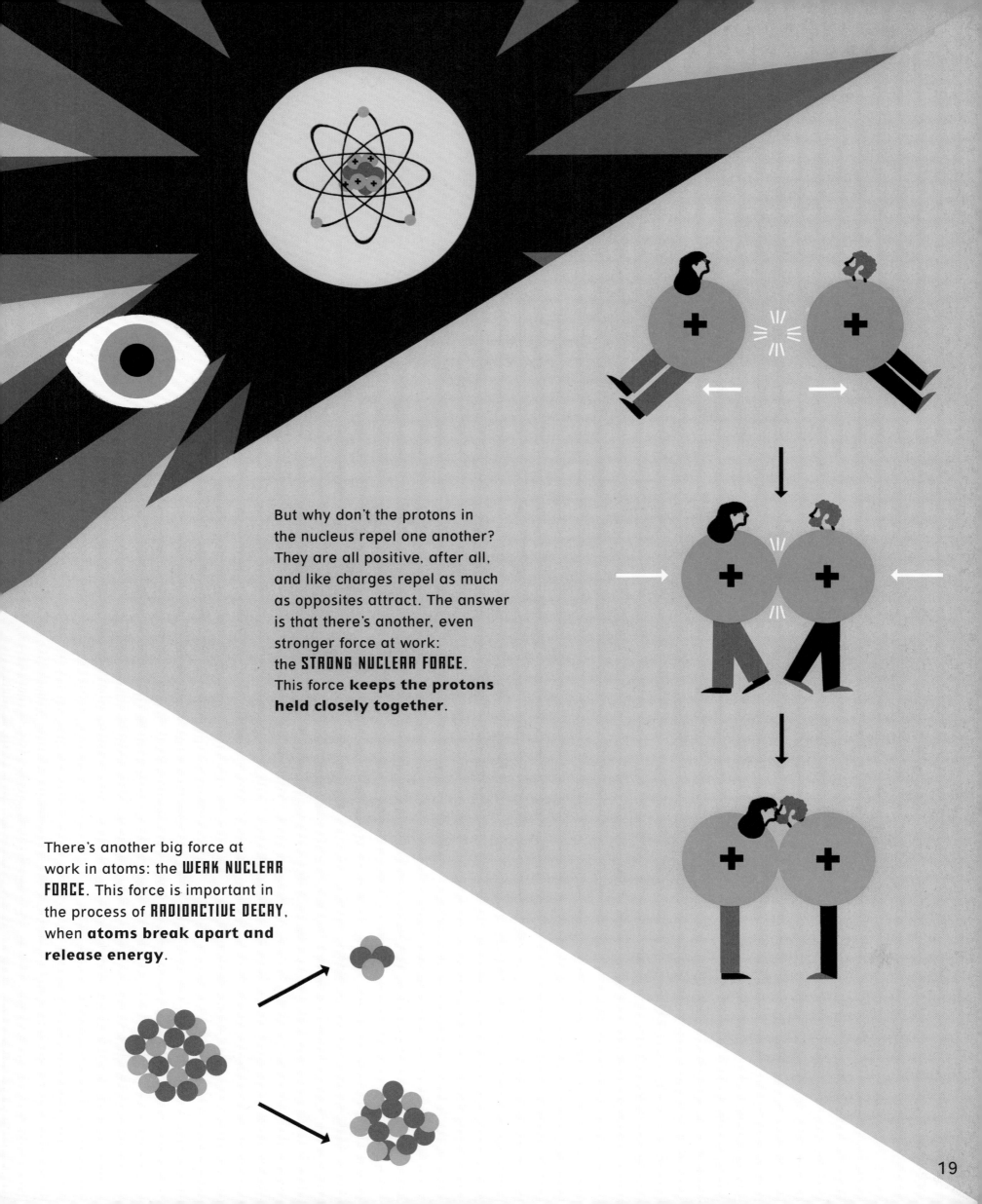

But why don't the protons in the nucleus repel one another? They are all positive, after all, and like charges repel as much as opposites attract. The answer is that there's another, even stronger force at work: the **STRONG NUCLEAR FORCE**. This force **keeps the protons held closely together**.

There's another big force at work in atoms: the **WEAK NUCLEAR FORCE**. This force is important in the process of **RADIOACTIVE DECAY**, when **atoms break apart and release energy**.

Energy

Energy is something that's needed to perform a task. When you eat food, your body turns some of it into energy to power your brain, heart, and other organs. Burning a candle turns the chemical energy in the wax into heat and light energy. A camera turns light energy into electrical energy.

Energy is also part of the most famous equation in all of physics:

$$E = mc^2$$

Here, **E is energy, m is mass,** and **c is the speed of light**. The brainchild of Albert Einstein, this equation says that energy and mass are essentially the same thing and that you can convert one into the other.

Energy is often transferred from one form to another, but it can never be destroyed. Physicists call this principle the **LAW OF CONSERVATION OF ENERGY**. All types of energy fall into one of two groups: kinetic or potential. Any object in **motion**, like a coasting bicycle, has **KINETIC** energy.

POTENTIAL ENERGY

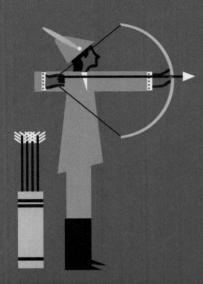

All objects have **POTENTIAL** energy. This is the energy that is **stored** within them. Drawing a bow increases its potential energy.

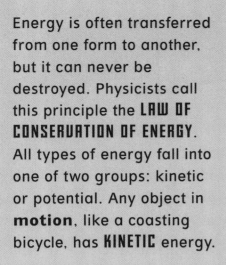

THE BIG BANG created a universe initially filled with only energy, but various processes have turned some of that energy into the matter we see around us in the form of stars, planets, and people.

21

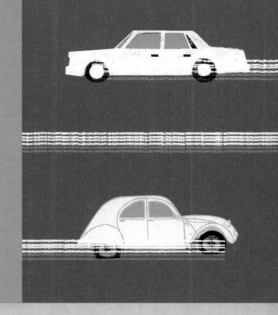

What Is Sound?

The world around us is a noisy place — birds singing, traffic rumbling, people talking, and music blasting. All these sounds are caused by vibrations in the air.

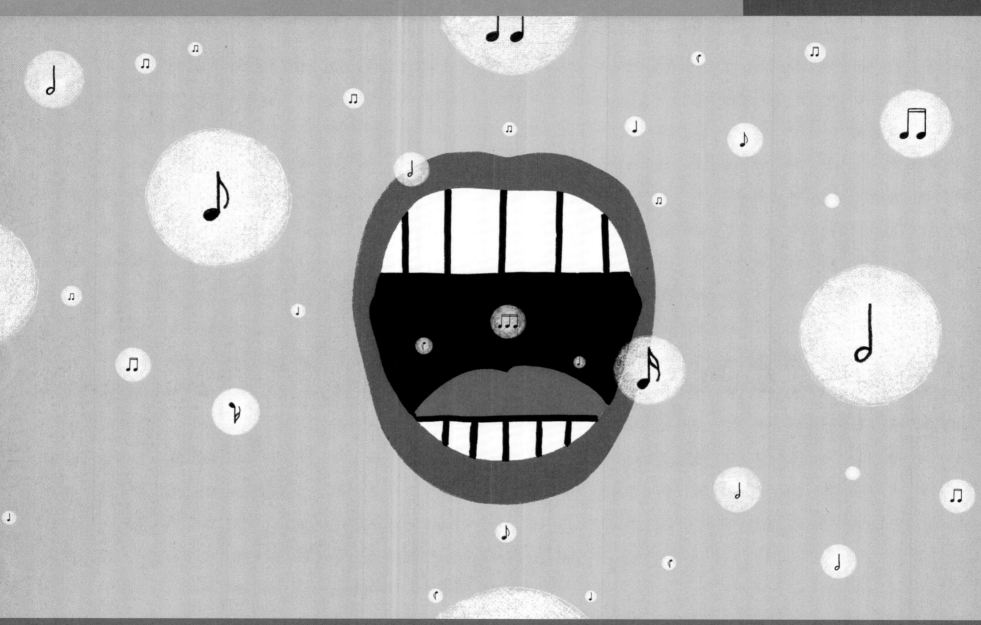

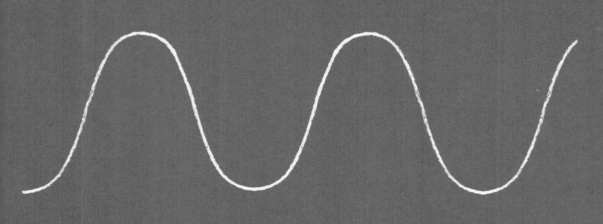

A speaker plays music by vibrating up and down, causing **AIR MOLECULES** nearby to **vibrate**, too. Then their neighbors vibrate, and then their neighbors, until eventually the air molecules next to your ears are also dancing. This is how sound travels as a wave through the air. These waves travel about a million times slower than light travels, which is why we always see lightning before we hear thunder.

How Do We Hear?

From beatboxing to Beethoven, we hear such rich and varied sounds thanks to our ears. The differences among sounds all come from the way microscopic hairs move deep inside our ears.

How the Ear Works:

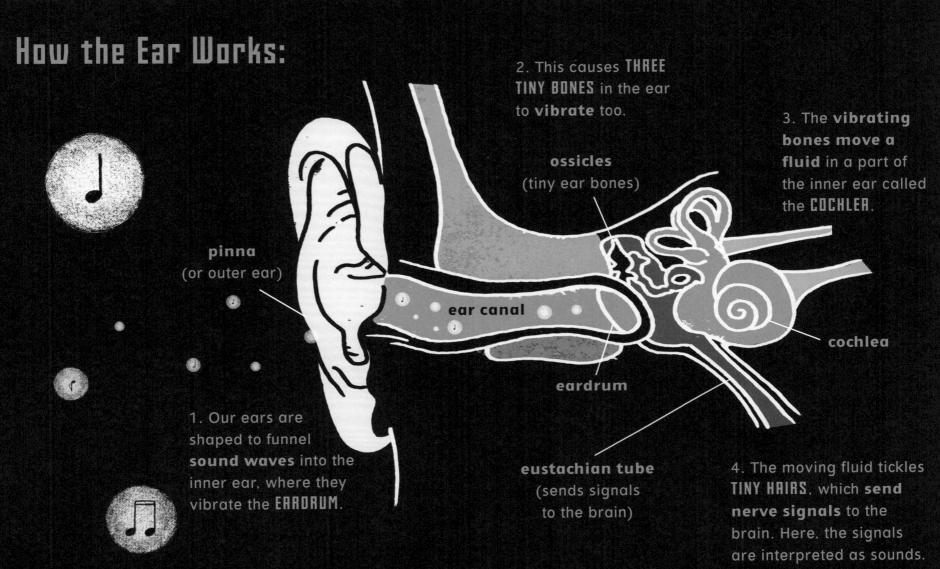

2. This causes **THREE TINY BONES** in the ear to **vibrate** too.

3. The **vibrating bones move a fluid** in a part of the inner ear called the **COCHLEA**.

ossicles
(tiny ear bones)

pinna
(or outer ear)

ear canal

cochlea

eardrum

1. Our ears are shaped to funnel **sound waves** into the inner ear, where they vibrate the **EARDRUM**.

eustachian tube
(sends signals to the brain)

4. The moving fluid tickles **TINY HAIRS**, which **send nerve signals** to the brain. Here, the signals are interpreted as sounds.

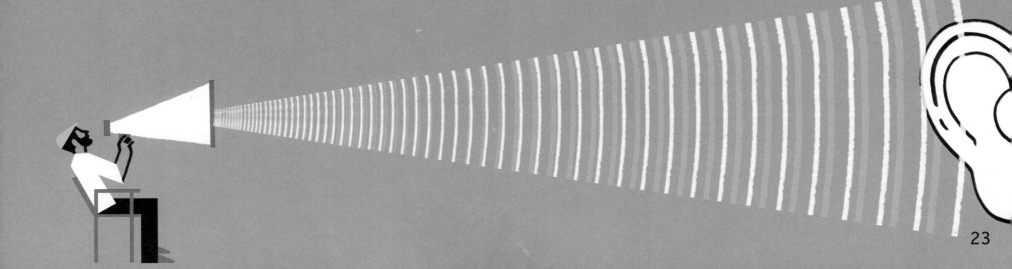

High and Low

There is a whole realm of sound above and below what we can hear. Some deep rumbles are **too low** to be heard [**INFRASONIC**], and other squeaks are too **high-pitched** [**ULTRASONIC**]. On average, humans can detect sounds that vibrate between 20 and 20,000 times a second. We measure the **frequency of sound vibrations** in **HERTZ** [**Hz**].

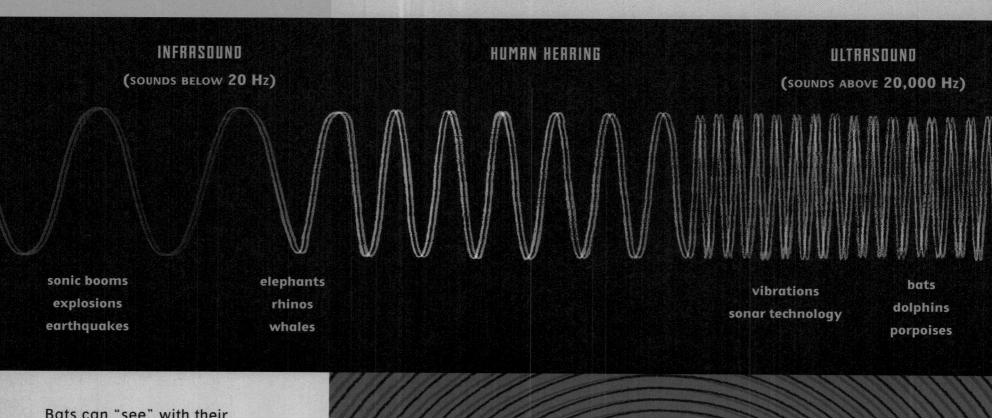

INFRASOUND
(SOUNDS BELOW **20 Hz**)

HUMAN HEARING

ULTRASOUND
(SOUNDS ABOVE **20,000 Hz**)

sonic booms
explosions
earthquakes

elephants
rhinos
whales

vibrations
sonar technology

bats
dolphins
porpoises

Bats can "see" with their **ULTRASONIC** ears better than humans see with their eyes in daylight. They hunt in total darkness by using **ECHOLOCATION**: they scan their environment with ultrasonic squeaks. Some of the sound, which **vibrates 110,000 times a second**, bounces back, and by swiveling their ears, bats can close in on their prey.

Certain insects that can detect these sounds have evolved ways to keep themselves from becoming dinner. Some moths, for example, close their wings and drop to the ground, while others fire out sonar-jamming clicks.

Because light scatters quickly underwater, it is hard to see long distances below the ocean's surface. Instead, whales communicate using a range of **INFRASONIC** grunts, groans, snorts, and barks. They also do something quite remarkable: they sing. Male humpback whales sing tunes that can be heard on the other side of the ocean. Mastering their songs may get them a whale girlfriend!

Toothed whales, such as dolphins and porpoises, also use ultrasound **ECHOLOCATION** to find prey. They focus **high-pitched clicks** made inside their skulls into a beam of sound. When this "sound ray" hits fish, part of it bounces back, allowing the hunters to pinpoint their prey. Humans have borrowed this trick to create sonar technology, which we use to map the seafloor and to find schools of fish.

Pitch Perfect

Some sounds are pleasing to hear, but others cause you to put your hands over your ears. What makes a piano nicer to listen to than the rumble of traffic? It is all to do with harmonies.

Both traffic and a piano solo are made up of several sounds of different frequencies (meaning that the sounds' waves are vibrating at different speeds). Some musical notes sound good when played together; others sound horrible. The ancient Greeks figured out that a note played with another that's double the frequency sounds harmonious. **Several harmonious notes** played together is known as a **CHORD**.

Musical instruments are able to create **NOTES** — **sounds with a constant frequency** — by making vibrations. Pluck a guitar string and it vibrates at a certain speed, or frequency. Press the string down against the guitar's fret board, and you shorten the string; it will vibrate faster, or at a higher frequency, making a higher note. Similarly, blowing into a wind instrument makes a sound by causing the air inside it to vibrate.

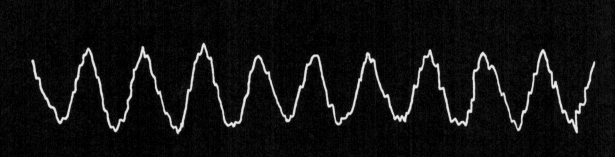

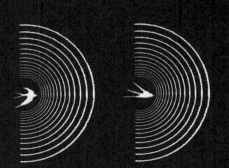

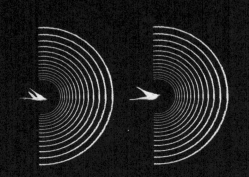

WHY DO BIRDS SING?

Have you ever listened to a birdcall? As well as sounding pleasant, bird songs are also very important for birds' survival, enabling them to locate a mate or warn of danger. Scientists have discovered that hearing is so important to birds that they can regenerate destroyed cells in their ears and repair hearing loss. Studying how birds do this may one day help us, too.

Technically Speaking

The range of human vocalization is amazing. We can make more than five hundred noises, and we can speak at different volumes, from a whisper to a scream.

There are nearly seven thousand different languages spoken around the world. Although our voice box allows us to speak, the teeth, tongue, and lips are also hugely important in making sounds. We combine them in different ways to create a whole multitude of noises. To test this, try saying the word *top* in front of a mirror. Watch how your teeth and mouth move to create the sounds. When you pronounce the letter *t*, your tongue touches the top of your mouth, stopping the flow of air, and when you pronounce the letter *p*, your lips press together.

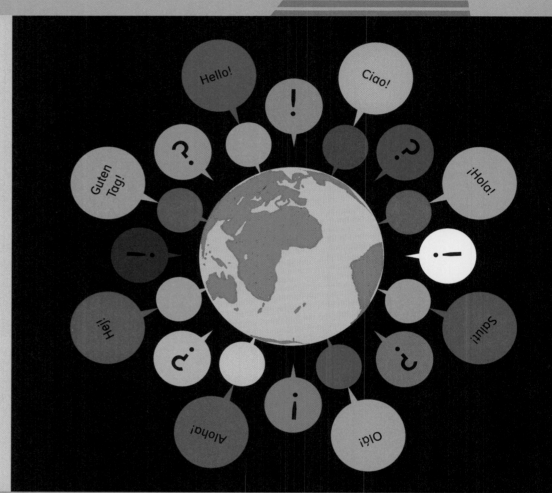

Human languages have so much variety! The Xhosa language, spoken in South Africa, uses clicks. Many Asian languages use rising or falling tones. And some communities, like the Hmong people of the Himalayas, also use a separate language that is whistled.

We can speak thanks to our **LARYNX**—or voice box—which sits at the top of our windpipe. If you touch your throat while you're talking, you can feel your larynx vibrating. It contains **folds of tissue** called **VOCAL CORDS** that vibrate as air passes through them. Men often have a pronounced "Adam's apple" visible in their throat—this is their voice box sticking out slightly.

WHY CAN'T ANIMALS SPEAK?

Although animals are capable of making a whole range of sounds, they cannot speak in the same way as humans. Among other reasons, it is thought that this is because they cannot control the air entering and exiting their lungs in the way humans can. Animals need to take more breaths between each noise they make. Instead, they use a whole range of body and facial movements to communicate alongside sound.

How Loud Is Loud?

We all know that some noises are louder than others — a cat purring is quieter than a drill, for example. Scientists measure the intensity of a sound in **decibels (dB)**. The unit gets its name from **ALEXANDER GRAHAM BELL**, the inventor of the telephone, and from the Latin *deci-*, meaning a tenth.

The quietest sound most people can hear is zero decibels. Each 10-decibel jump up the scale represents a sound ten times louder than the last. So a 20-decibel sound is one hundred (10 x 10) times louder than a zero-decibel sound, and a 40-decibel sound is 10,000 (10 x 10 x 10 x 10) times louder. Mathematicians call this type of progression a **logarithmic scale**.

rocket launching
180 dB

jet plane taking off
140 dB

live rock band
120 dB

The human ear is an amazing instrument capable of hearing sounds over a huge range, from rustling leaves (20 decibels) to a rock concert (120 decibels). Once sounds get to 130 decibels, they start to hurt your ears. At 160 decibels, they could well pierce your eardrum. This can lead to hearing loss, although a tear often repairs itself within a few weeks if left to heal.

motorcycle
100 dB

alarm clock
80 dB

conversation
60 dB

birdcalls
40 dB

rustling leaves
20 dB

pin drop
10 dB

Peace and Quiet

Imagine a place so quiet that you can hear the bones click in your joints, the blood coursing around your body, and even your eyes moving in your skull. A place where even swallowing is loud. This is exactly what happens inside anechoic chambers, the quietest places on the planet. The term **ANECHOIC** means "**having no echo**."

There's an anechoic chamber at Microsoft's headquarters in Redmond, Washington. Its sound level is an unimaginable −20.6 decibels—that's 100,000 times quieter than a whisper and way below what our ears can detect. It's so quiet that occupants can hear sounds from inside their own body that are normally drowned out.

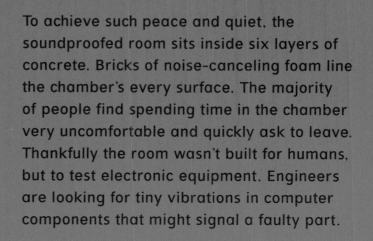

To achieve such peace and quiet, the soundproofed room sits inside six layers of concrete. Bricks of noise-canceling foam line the chamber's every surface. The majority of people find spending time in the chamber very uncomfortable and quickly ask to leave. Thankfully the room wasn't built for humans, but to test electronic equipment. Engineers are looking for tiny vibrations in computer components that might signal a faulty part.

It's impossible to find any natural spots on Earth that come close to the silence of the anechoic* chamber, but there are still a few remote places in the world that have no sounds made by humans.

The Hoh Rain Forest, in Washington State, is home to possibly the quietest outdoor place in the United States: deep within the forest, a small red stone marks "One Square Inch of Silence."

The Negev Desert, in Israel, is reportedly so silent that you can hear your ears ring and the sand sing in the scorching heat.

Breaking the Sound Barrier

Just as a boat makes ripples as it moves through water, an object traveling through air creates waves around it. These waves travel at the **SPEED OF SOUND**, which is around **1,125 feet/343 meters per second**.

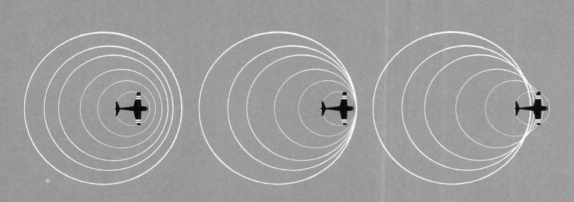

As the object itself approaches the speed of sound, the waves in front of it increasingly bunch up until they merge into **one giant shock wave** called a **SONIC BOOM**. For this reason, you cannot hear an object approaching you that is traveling faster than the speed of sound.

Humans are fascinated with making things go faster. Not long after the invention of planes, work began to make them fly even quicker. In 1947, the American test pilot Chuck Yeager became the first person to break the sound barrier, in a plane called the Bell X-1. He flew at 700 miles/1,127 kilometers per hour! We call the speed of sound **Mach 1**, after Austrian physicist **ERNST MACH**, who studied shock waves.

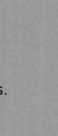

Since then, many planes have been able to fly even faster, thanks to improvements in design. Between 1976 and 2003, Air France and British Airways operated the Concorde, a supersonic airliner that could travel from London to New York in under three hours at more than twice the speed of sound (Mach 2). The current air-speed record is held by an American spy plane that can fly at 3.5 times the speed of sound!

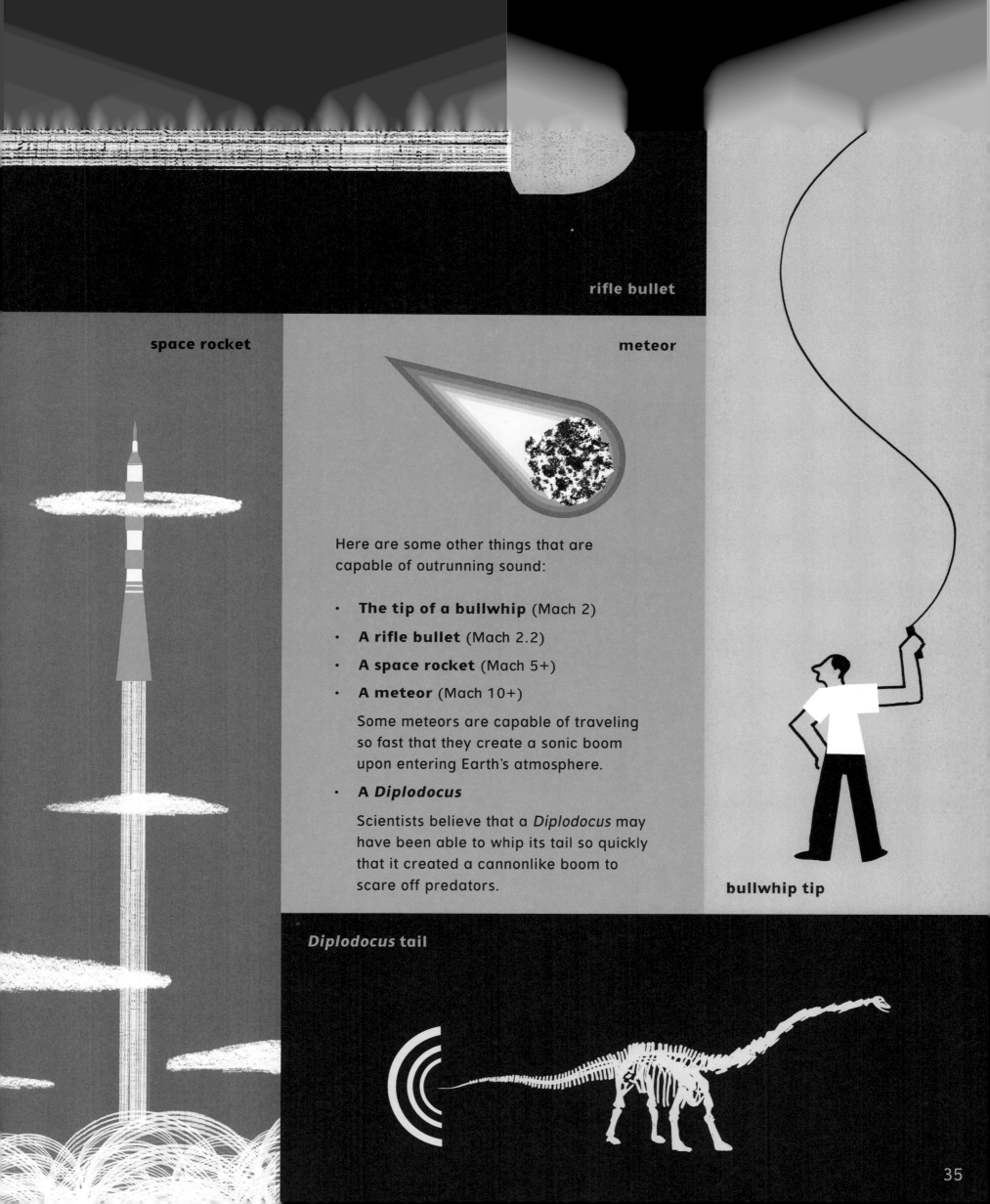

space rocket

rifle bullet

meteor

Here are some other things that are capable of outrunning sound:

- **The tip of a bullwhip** (Mach 2)

- **A rifle bullet** (Mach 2.2)

- **A space rocket** (Mach 5+)

- **A meteor** (Mach 10+)

   Some meteors are capable of traveling so fast that they create a sonic boom upon entering Earth's atmosphere.

- **A *Diplodocus***

   Scientists believe that a *Diplodocus* may have been able to whip its tail so quickly that it created a cannonlike boom to scare off predators.

bullwhip tip

Diplodocus tail

Seismic Shakes

Suddenly, without warning, the ground starts to shake. Books fly off shelves as you huddle in a doorway for shelter. You're caught in the middle of an earthquake.

Scientists who study earthquakes are called **SEISMOLOGISTS**. They use seismic waves to understand what's going on deep inside the Earth and measure the strength of an earthquake on the **Richter scale**. The scale was developed by **CHARLES RICHTER** in 1935 and uses **numbers to represent the amount of energy released by earthquakes**. Each jump up the scale is 10 times more powerful than the last.

The Richter Scale

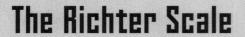

9 GREAT
Rare, but near-total destruction of buildings, roads, and bridges.

8 GREAT
Major destruction. Buildings and structures may collapse; bridges may be destroyed.

7 MAJOR
Quake can be detected all over the world.

6 MODERATE
Great damage around the epicenter. Cracks in the ground may appear, and underground pipes may burst.

5 MODERATE
Damage caused to weaker buildings near the epicenter. Furniture may move, and plaster could loosen from walls and ceilings.

4 SMALL
Some damage to buildings near the epicenter, including broken windows.

3 SMALL
People near the epicenter may feel this quake as vibrations.

2 MINOR
Smallest quake felt by people. Some hanging objects may swing.

1 INSIGNIFICANT
Humans cannot detect these quakes, which happen almost daily.

These dramatic and violent events happen because the Earth's crust is not one solid surface. Instead, it is made up of a series of **interlocking jigsaw puzzle pieces** called **TECTONIC PLATES**. They float on top of a hot ocean of molten rock called magma. The Earth quakes when two plates rub against each other.

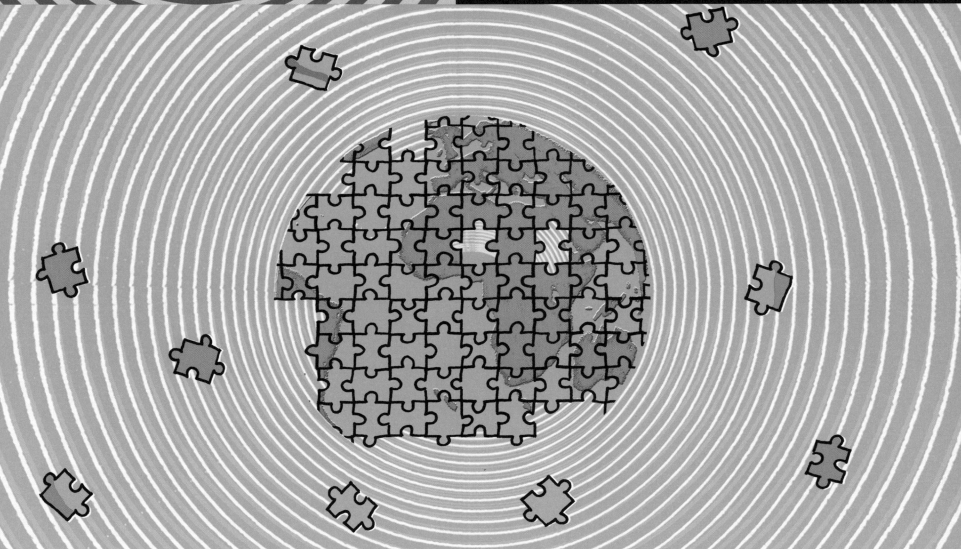

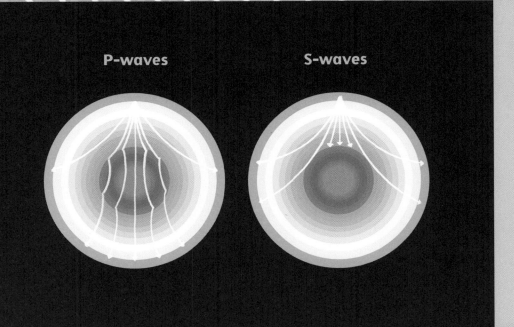

P-waves   S-waves

The energy from an earthquake travels through the Earth in the form of waves. Scientists split the waves into two types, called **P-WAVES** and **S-WAVES** (pressure waves and shear waves). They **travel through the planet at different speeds and in different ways**.

P-waves are similar to sound waves. They can travel through both solids and liquids and all the way through the Earth. S-waves can only travel through solids, so they cannot pass through the Earth's molten outer core.

Noises of the World

Here are some of the loudest and strangest sounds ever recorded on Earth.

North
America

HOWLER MONKEY: 90 dB

The unmistakable call of these noisy primates is generated by the hyoid bone in their throat.

THE BLOOP, PACIFIC OCEAN: 0–50 Hz

During the 1990s, a strange noise picked up in the Pacific Ocean was hailed as one of the loudest sounds ever recorded underwater. Some scientists think it was caused by vibrations from underwater volcanoes, but the sound disappeared before we could figure out what caused it. It is known as the Bloop.

Pacific
Ocean

South
America

GREATER BULLDOG BAT: 140 dB

Native to South America, these bats are also known as fisherman bats because they use loud sounds to locate fish.

IGUAZU FALLS, BRAZIL AND ARGENTINA: 100 dB

Iguazu Falls are the largest waterfall system in the world, with water pouring over cliffs up to 269 feet (82 meters) high.

Southern
Ocean

Arctic Ocean

Asia

Europe

Africa

Pacific Ocean

Oceania

Antarctica

TUNGUSKA, SIBERIA: 300 dB

In 1908, a meteor exploded above the ground in northern Russia. Luckily the area is deserted as the noise reached 300 decibels.

PISTOL SHRIMP: 200 dB

These feisty creatures stun their prey with powerful jets of water from their claws, creating a shockwave louder than a gunshot.

KRAKATOA, INDONESIA: 172 dB

In 1883, this volcano erupted with such force that someone 100 miles/160 kilometers away would have heard an enormous bang. Some say it was the loudest sound ever heard on Earth.

LION: 114 dB

As you may expect, the mighty roar of a lion is pretty loud and can be heard from 5 miles/ 8 kilometers away.

ICEBERG, ANTARCTICA: 220 dB

An "icequake" occurs when an iceberg calves off a glacier. The resulting crack and boom can be louder than a rocket launching.

AUSTRALIA: 50–200 Hz

In the waters around Australia, a strange quacking sound has been heard on and off since the 1960s. Locals have called it "bio-duck." Scientists' best guess is that it is made by minke whales as they come up to the surface for air.

39

Sounds in Space

Sound cannot travel through space because there is no air to carry the vibrations. It's true what they say — in space, no one can hear you scream!

Astronomers *can* look at vibrating objects throughout the universe and turn those vibrations into sound to help them understand what's going on. Stars like the sun throb and pulse, and **astronomers can use these vibrations to look deep inside** its layers. This is called ASTEROSEISMOLOGY.

As two black holes spin around each other, they **create vibrations** in the very fabric of space itself. Astronomers call these vibrations GRAVITATIONAL WAVES. The frequency of the waves increases as the black holes get nearer to each other. When the black holes collide, they produce a distinctive "chirp." Physicists have translated the chirp into frequencies the human ear can hear.

The lowest note we've found in the universe is produced by a vibrating black hole in the Perseus cluster. It is a million billion times lower than the lowest note your ears can hear and would sit 57 octaves below middle C on a piano keyboard. The left-hand end of a piano would have to be extended by 30 feet/ 9 meters in order to play it!

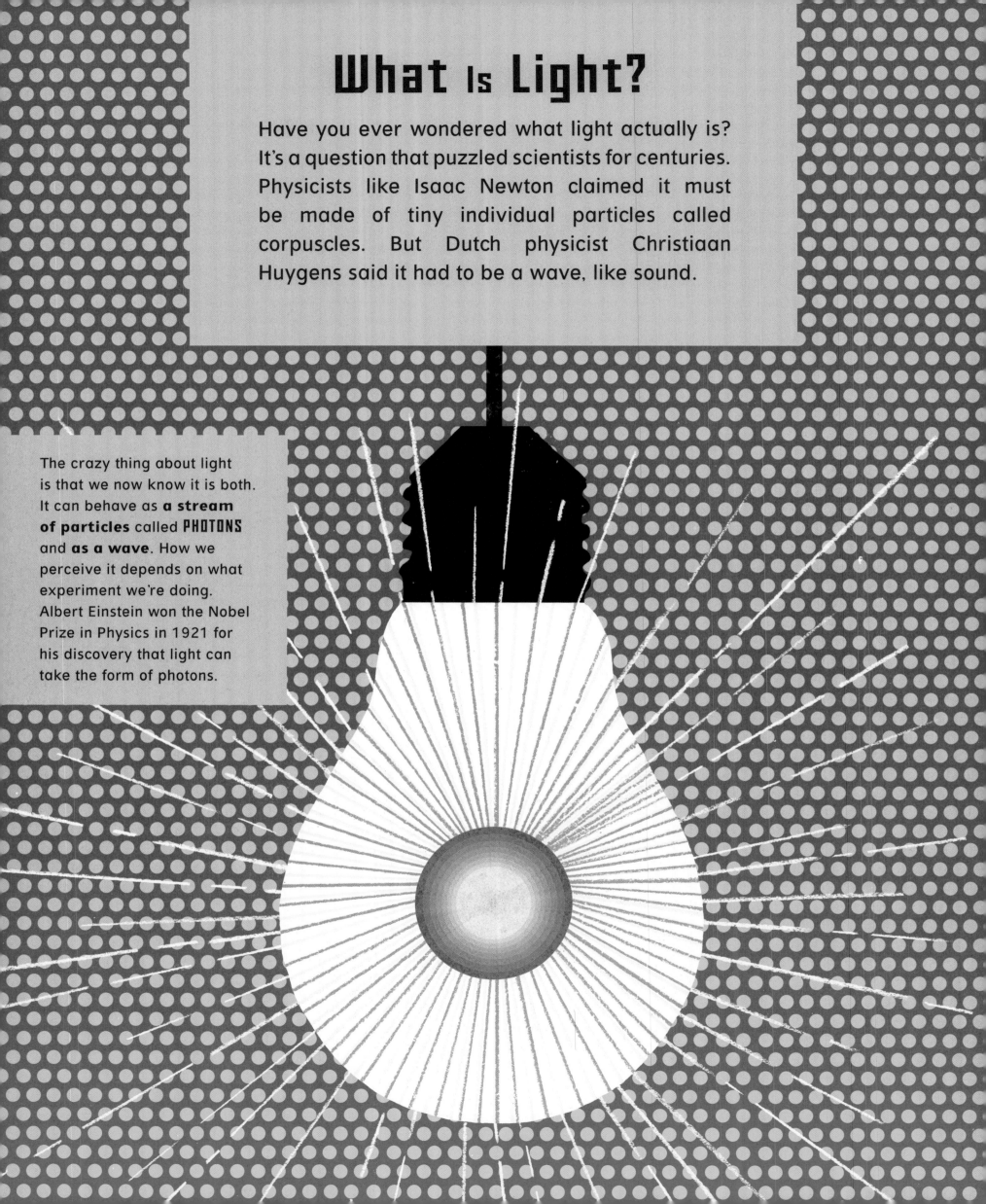

What Is Light?

Have you ever wondered what light actually is? It's a question that puzzled scientists for centuries. Physicists like Isaac Newton claimed it must be made of tiny individual particles called corpuscles. But Dutch physicist Christiaan Huygens said it had to be a wave, like sound.

The crazy thing about light is that we now know it is both. It can behave as **a stream of particles** called PHOTONS and **as a wave**. How we perceive it depends on what experiment we're doing. Albert Einstein won the Nobel Prize in Physics in 1921 for his discovery that light can take the form of photons.

How Do We See?

We are able to see the world around us because light reflects off nearby objects and into our eyes.

How the Eye Works:

1. First, light travels through the **CORNEA**—**the thin transparent** layer at the front of the eye—into the **PUPIL**.

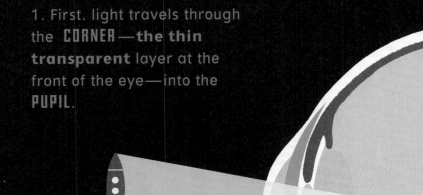
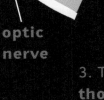

cornea

pupil

lens

retina

optic nerve

2. Next, a **LENS focuses the light** onto a **projection screen** at the back of the eye called the **RETINA**, where the image is turned into electrical signals.

3. The **OPTIC NERVE** sends those signals to the **brain** to make sense of.

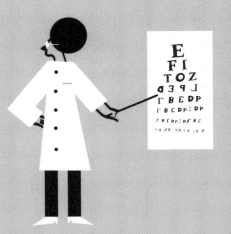

The Speed of Starlight

Light moves incredibly fast. Flick on a light switch and—BOOM—it instantly fills the room. As far as we know, light is the fastest thing in the universe, but it is not infinitely fast.

Light travels at nearly 200,000 miles/ 300,000 kilometers per second. That's about a million times faster than sound and means that light can whiz around the Earth seven and a half times in one second. A **LIGHT-YEAR** is **the distance light travels in a year.** This works

out at about 5.9 trillion miles/ 9.5 trillion kilometers and is a good way to measure the astronomical distances in space. Proxima Centauri, the star that's closest to us after the sun, is just over four light-years away, which means its light takes four

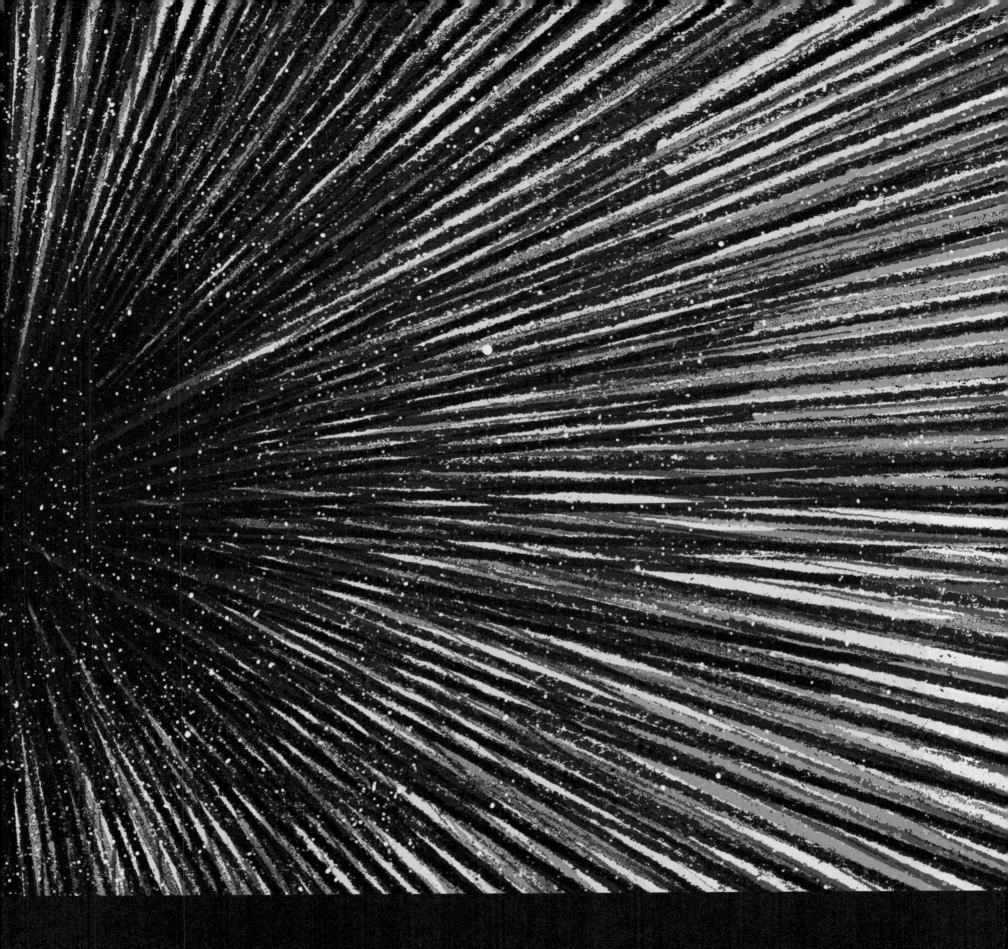

years to reach Earth. Its light takes 27,700 years to reach the center of our galaxy, the Milky Way; 2.5 million years to reach Andromeda (our nearest large galaxy); and a colossal 46.5 billion years to reach the farthest galaxies in the universe.

SPEED OF LIGHT =
983,571,056 feet per second
Nine hundred eighty-three million, five hundred seventy-one thousand, fifty-six feet per second

ONE LIGHT YEAR =
31,039,141,970,409,448 feet
Thirty-one quadrillion, thirty-nine trillion, one hundred forty-one billion, nine hundred seventy million, four hundred nine thousand, four hundred forty-eight feet

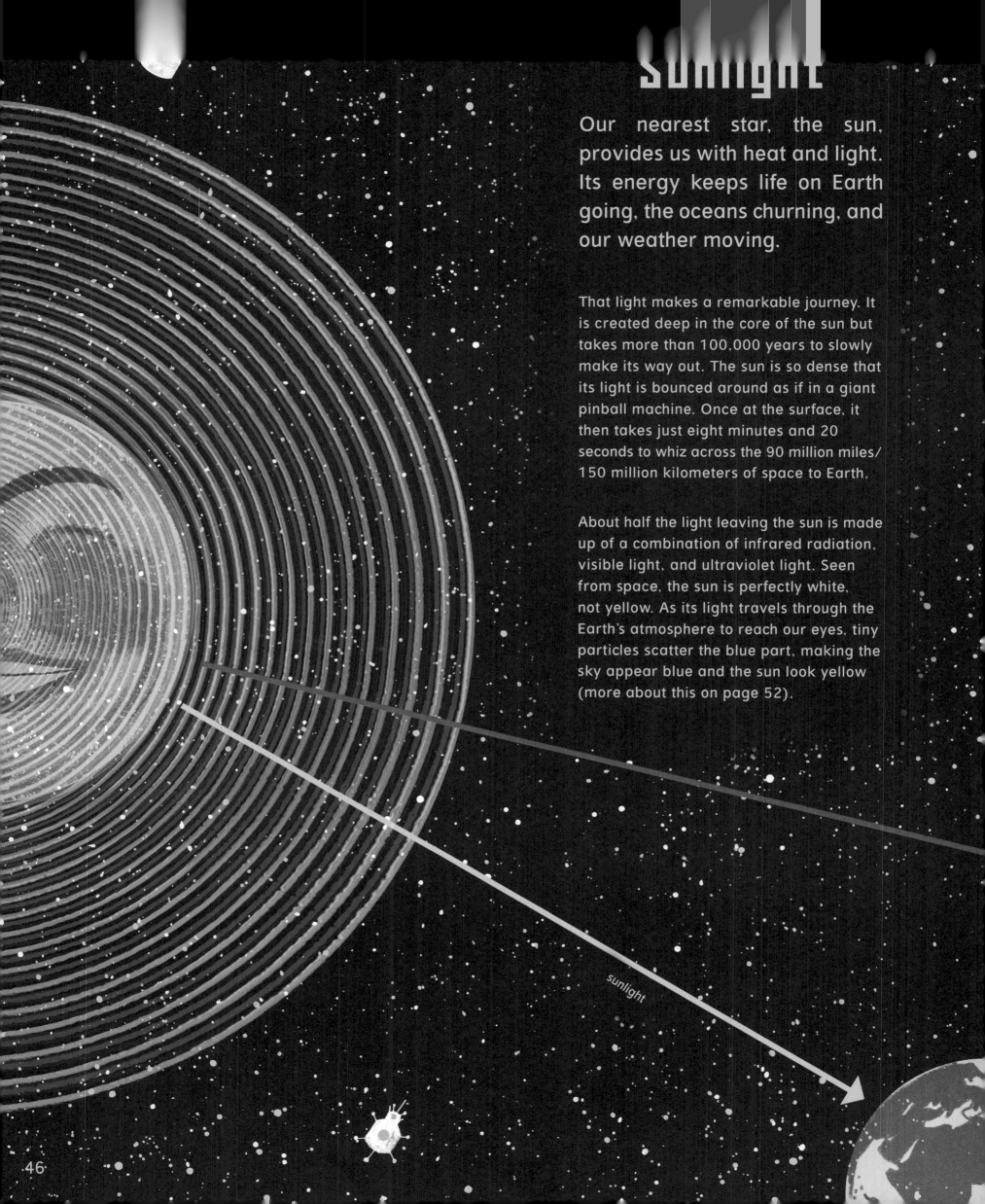

Sunlight

Our nearest star, the sun, provides us with heat and light. Its energy keeps life on Earth going, the oceans churning, and our weather moving.

That light makes a remarkable journey. It is created deep in the core of the sun but takes more than 100,000 years to slowly make its way out. The sun is so dense that its light is bounced around as if in a giant pinball machine. Once at the surface, it then takes just eight minutes and 20 seconds to whiz across the 90 million miles/ 150 million kilometers of space to Earth.

About half the light leaving the sun is made up of a combination of infrared radiation, visible light, and ultraviolet light. Seen from space, the sun is perfectly white, not yellow. As its light travels through the Earth's atmosphere to reach our eyes, tiny particles scatter the blue part, making the sky appear blue and the sun look yellow (more about this on page 52).

sunlight

Moonlight

On a bright, moonlit night, the moon looks as if it is shining with a pale, silver light. But the moon does not glow — its light is reflected sunlight and even "earthshine" bouncing off its surface, exactly like a mirror.

The moon is actually a pretty terrible mirror. Its surface is bumpy and dark gray. So, despite being the brightest object in our night sky, the moon only reflects just over a tenth of the light that hits it.

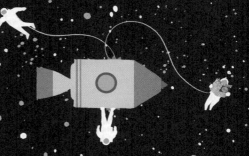

Sun's light reflected as "moonlight"

Apollo astronauts left reflectors on the moon, which can bounce light fired from lasers back to Earth. Carefully measuring the time taken for the light to return has shown that the moon is drifting 1.5 inches/3.8 centimeters farther away from Earth every year—that's about as fast as your fingernails grow.

The Sun's Core

The sun is the only object in our solar system capable of making light on its own.

Every hour, the sun fuses 2.46 trillion tons/2.23 trillion metric tons of protons (hydrogen) into 2.45 trillion tons/2.22 trillion metric tons of helium. The missing 15 billion tons/ 14 billion metric tons are turned into sunlight that pours out of the core. Even at this crazy rate, it will take the sun billions more years to exhaust its supplies.

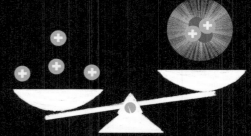

Sunlight is made deep in the heart of the sun, where gas is squashed together so fiercely that it becomes thirteen times denser than lead. Temperatures in the core soar to a colossal 27 million degrees Fahrenheit/15 million degrees Celsius, as every second, trillions upon **trillions of protons are forced together to create helium**. Astronomers call this process **FUSION**.

Scientists have been trying to copy the sun to make a fusion power station on Earth. The machines they use are called **TOKAMAKS**, and temperatures inside can reach **180 million degrees Fahrenheit/100 million degrees Celsius**—hotter than the sun's core.

Ferocious Flashes

If you think the sun's power is immense, it's nothing compared to the ferocious fury of a **GAMMA-RAY BURST [GRB]**.

In less than a minute, a GRB can release as much energy as the sun will over its ten-billion-year lifetime. These brilliant flashes are created by either an exploding star or the cores of two stars smashing together after a dizzying death spiral.

GRBs belong to a group of objects studied by high-energy astronomers, who explore the short-lived displays associated with black holes, neutron stars, and supernova explosions—the universe at its most extreme.

Some objects can get so bright that they can be seen more than halfway across the universe, at distances where normal stars would be far too faint. In the middle of old galaxies, gargantuan black holes gorge on huge clouds of stars and gas, spewing out **vast quantities of X-rays and gamma rays**. Astronomers know these phenomena as **QUASARS** and **BLAZARS**, and they can shine brighter than a thousand galaxies combined, even though a galaxy is made up of hundreds of billions of stars.

Making Food from Sunlight

Eating is one of the best things about being human—we have so many delicious things to choose from. But it can also be a bother to have to remember to eat all the time. What if we could use our bodies to make our own food from scratch, whenever we wanted? Well, that's exactly what plants do. It is called **PHOTOSYNTHESIS**.

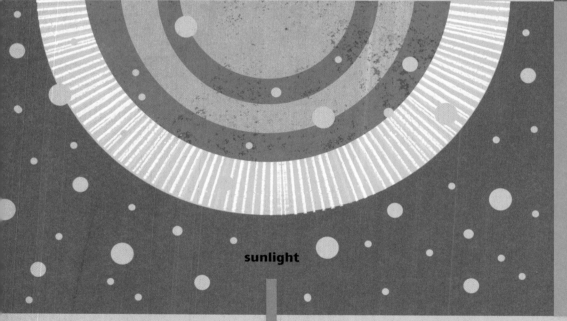

sunlight

When you eat fruit and vegetables, the energy from their sugar passes on to you. So energy moves from the sun to plants, then animals, and then you in a series of steps that scientists call a **FOOD CHAIN**. Many **overlapping food chains in an ecosystem are known as a FOOD WEB**. The first link in a chain is always the sun—so even when you eat a hamburger, what you're really eating is repackaged solar energy!

plant

cow

human

Inside a Plant Cell

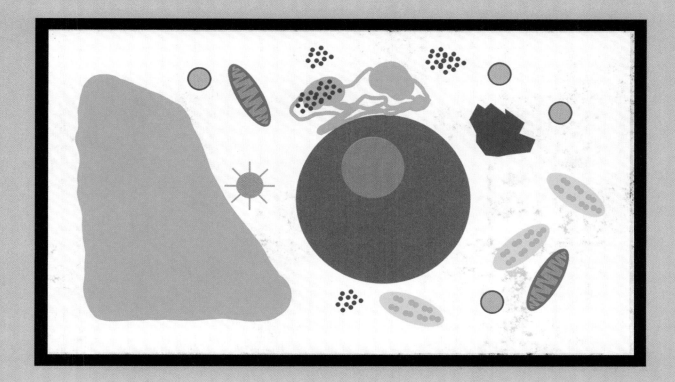

Plants use a green chemical called **CHLOROPHYLL** to **absorb sunlight** with their leaves. They combine this with water and carbon dioxide from the air to create oxygen and sugar. They use the sugar as food and release the oxygen into the atmosphere.

Photosynthesis

The **process by which plants create food and oxygen** is called **PHOTOSYNTHESIS**.

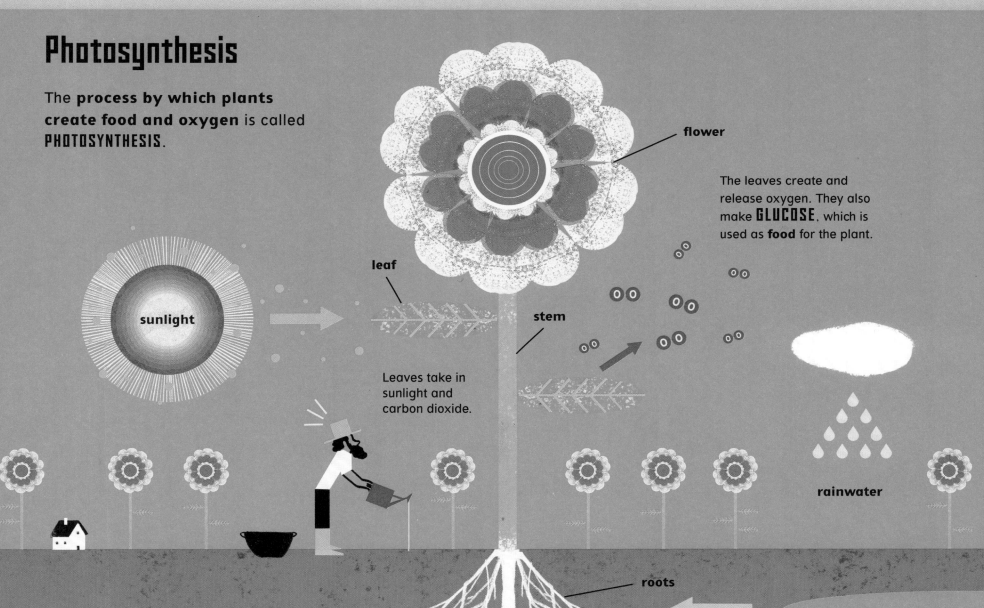

flower

The leaves create and release oxygen. They also make **GLUCOSE**, which is used as **food** for the plant.

leaf

Leaves take in sunlight and carbon dioxide.

sunlight

stem

rainwater

roots

Roots absorb water and nutrients from the soil.

Why Is the Sky Blue?

You can think of light as a wave, much like a wave on the ocean. The **distance over which a wave repeats** is called its WAVELENGTH. Red light has a much longer wavelength than blue light.

The different colors of sunlight are scattered by air molecules as they enter Earth's atmosphere. How much a color gets scattered depends on its wavelength.

52

Imagine that light is taking footsteps across the sky. **RED LIGHT**, with its **long wavelength**, has big footsteps like a giant so it can easily walk over most of the air molecules.

However, **BLUE LIGHT** can only take much **shorter** steps and so it gets scattered the most. When you look up at the sky in the middle of a clear day, most of the scattered light that hits your eyes is blue.

At sunrise and sunset, however, the light from the sun has to travel through a lot more atmosphere to reach our eyes. Only the light with the biggest footsteps—the longest wavelength—can make it through all that gas. That's why the sky can appear red or reddish at these times.

Splitting
the
Rainbow

With the smell of rain still in the air, the sun emerges from the clouds and a glorious rainbow stretches across the sky. These vivid, colorful arches are caused by sunlight entering water droplets.

You can only see a rainbow if the sun is behind you because the light goes in front of the raindrop, hits the back, and then returns toward you.

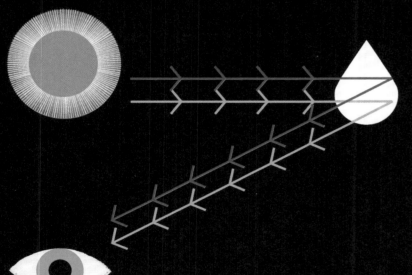

This journey through the water droplet splits white light up into its seven colors: red, orange, yellow, green, blue, indigo, and violet.

There is often a second, fainter rainbow, too. Look closely and you'll spot that its colors are reversed. The sky is also noticeably darker between the two rainbows. This **dingy patch** is called ALEXANDER'S DARK BAND.

We're so used to seeing rainbows as arcs, but that's only because the ground normally blocks out the rest of the circle. Mountaineers and pilots can sometimes see a complete, circular rainbow. Earth isn't the only planet with rainbows, either—astronomers have spotted one on Venus and think rainbows might also occur on Saturn's largest moon, Titan.

Spectacular Eclipses

The night sky is full of beautiful objects, but few events rival the jaw-dropping spectacle of a **SOLAR ECLIPSE**. The **moon creeps in front of the sun, blocking out most of its light** and causing temperatures to drop. Animals that were busy going about their business go quiet in confusion, creating an eerie silence.

For many people, the most impressive part of a solar eclipse is the **DIAMOND RING EFFECT**. The edge of the moon is not perfectly smooth—it has valleys and mountains that make the rim jagged. If light spills through one of these gaps, it looks like a precious jewel.

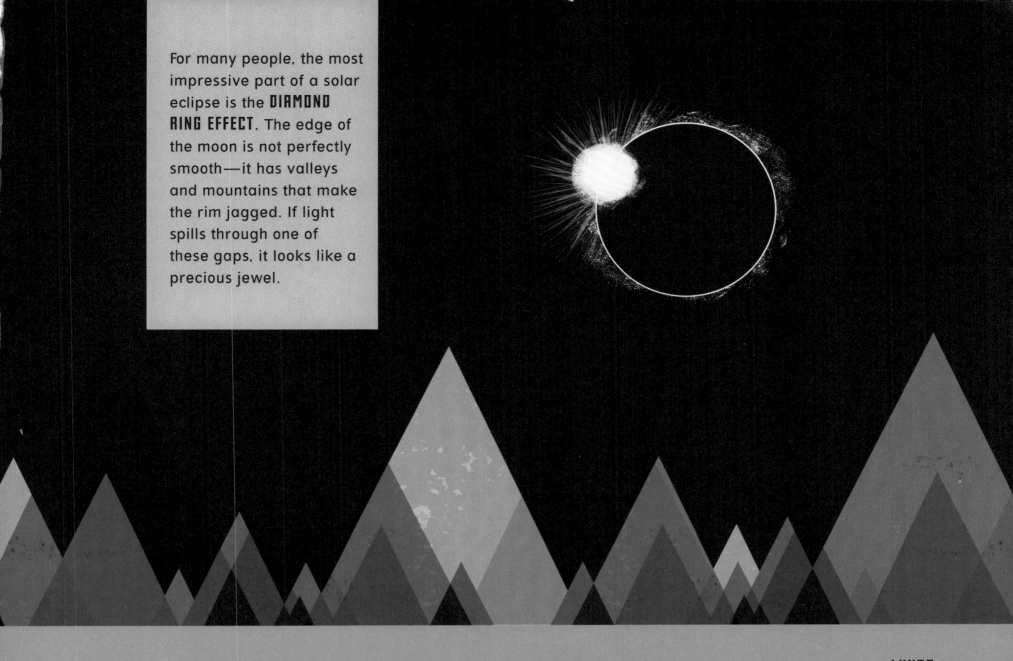

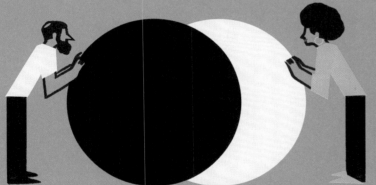

The other main type of eclipse is a **LUNAR ECLIPSE**. These occur when **the Earth is between the sun and the moon**. Normally this would stop any sunlight from reaching the moon at all. But the atmosphere above our heads is able to bend the red part of the color spectrum around our planet. This effect turns the moon a spooky, deep blood-red.

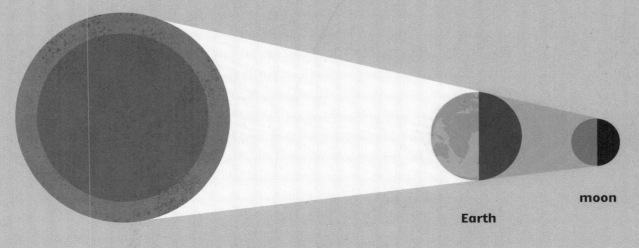

sun

Earth

moon

Nature's Light Displays

The natural world is capable of putting on displays of light as spectacular as any fireworks display. People living close to Earth's poles are often treated to the beauty of **AURORAE**, also known as the **northern** and **southern lights**. They are caused when atoms in our atmosphere receive extra energy from space and give out light. Clapping, clicking, and popping noises can also be heard as these curtains of light dance overhead.

Several times a year, we are treated to glorious **METEOR SHOWERS**—bursts of "shooting stars" that tear across the sky. They are actually not stars at all, but **tiny grains of space dust** burning up high in our atmosphere.

Creatures in the ocean such as jellies and squid are good at making impressive light shows as well. This is called **BIOLUMINESCENCE.** Glowing algae once saved the life of astronaut Jim Lovell, the commander of the Apollo 13 mission, when he was a fighter pilot. His night navigation equipment failed, and he couldn't find his aircraft carrier—that is until he saw the bioluminescent algae being churned up by the ship. He followed the glowing water all the way to a safe landing.

Telescopes

One way to think about telescopes is to picture them as giant light buckets. Imagine that the starlight falling to Earth is rain. If you want to collect more rain, you need a bigger bucket. Since our eyes are pretty small, they can only collect so much light on their own. That's why we build bigger buckets (telescopes). They enable us to see faint-looking objects that are far away.

There are two main types of telescopes: **REFLECTORS** (which use **mirrors**) and **REFRACTORS** (which use **lenses**). The biggest telescopes in the world are all reflectors, as it is easier and cheaper to build big mirrors.

A telescope's location is very important. It needs to be in a place that tends to have clear skies so there aren't too many cloudy nights. And since the atmosphere can make stars blurry, it also needs to be somewhere high up so that there is as little air as possible to look through. One of the world's best telescopes, the wonderfully named **VERY LARGE TELESCOPE [VLT]**, sits 8,645 feet/2,600 meters above sea level in the bone-dry Atacama Desert, in Chile.

Invisible Rays

Just as there are sounds too low and high for our ears to hear, there is light too low or high in frequency for us to see. In fact, most of the light around us is invisible. Scientists have invented some clever ways for us to see what our eyes cannot.

Below the red end of the color spectrum—with increasingly lower frequencies and increasingly longer wavelengths—is infrared light, then microwaves and radio waves. Above the violet end—with increasingly higher frequencies and increasingly shorter wavelengths—is ultraviolet light, then X-rays and gamma rays. The **complete set of light rays** is called the ELECTROMAGNETIC SPECTRUM.

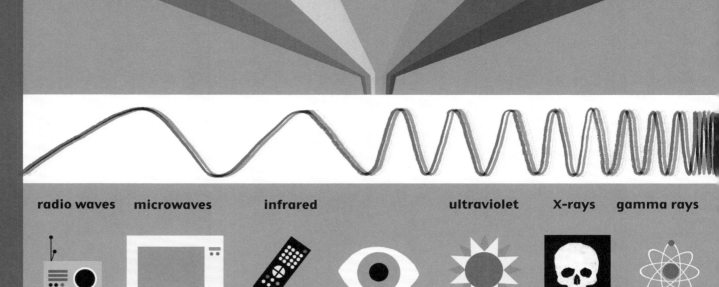

radio waves **microwaves** **infrared** **ultraviolet** **X-rays** **gamma rays**

The universe is playing a symphony using all these invisible rays, but our eyes can pick up on only a few notes. Black holes churn out X-rays, colliding neutron stars flash bright in gamma rays, and pulsating stars beam out radio waves. If we restricted ourselves to just the light our eyes can see, we'd really be missing out, so we build telescopes to see the light we can't.

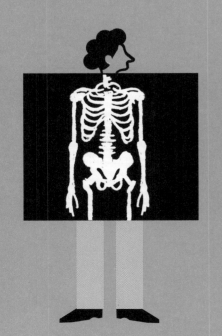

Some parts of the electromagnetic spectrum, such as X-rays, are emitted from celestial objects but don't make it to Earth, so we have to launch telescopes into space if we want to detect them. Only visible light and radio waves make it down here through our atmosphere and magnetic field.

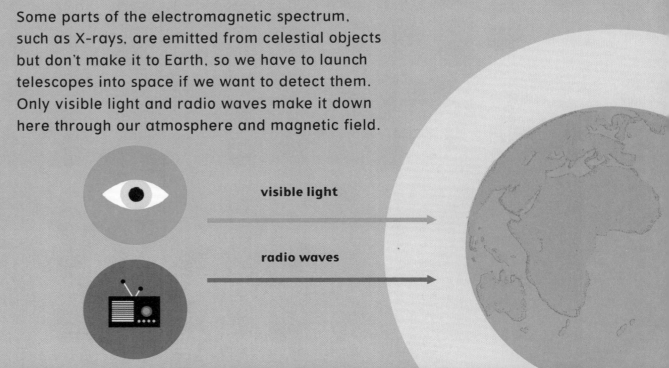

visible light

radio waves

In the Beginning

The **oldest light in the universe** is called the COSMIC MICROWAVE BACKGROUND [CMB]. To start with, the universe was too jam-packed for light to travel very far. But, 380,000 years after the big bang, the expansion of the universe meant there was suddenly enough space for light to flood out freely. Scientists have discovered microwaves flowing from every direction in space; they believe they all came from one source.

That makes the CMB the equivalent of the universe's baby picture. If the universe were a forty-year-old, the CMB is a snapshot of what it was like when it was just ten hours old!

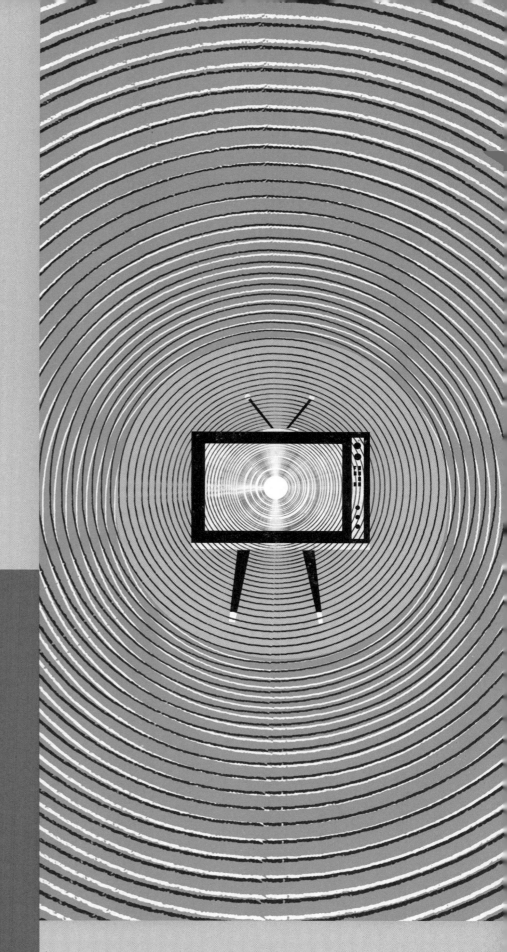

The CMB was originally discovered by accident in the 1960s by two American astronomers, **Arno Penzias** and **Robert Wilson**. They detected a hum in their radio telescope that they couldn't get rid of. At one point they thought it might be due to poop from pigeons roosting in their antenna, but the signal remained even after the pigeons had been evicted. What they were really picking up was the **afterglow of the big bang.**

You, too, can tune into the CMB—with just an analog TV or radio. One percent of the crackling interference you get between stations is due to this ancient light from the birth of our universe.

Searching for Aliens

It's arguably the biggest of all questions: Are we alone in the universe? To find out, astronomers have been scanning space, searching for planets around other stars.

Planets in other solar systems are too small, dim, and far away to be seen directly. Instead, astronomers have clever ways to tell if one is there. Scientists don't just *look* for aliens; they also listen out for them with radio telescopes. For decades they have been scanning the skies for any messages from ET, but so far they haven't heard anything.

If a planet moves in front of its star, then we see the star get a little dimmer. Planets also have a small gravitational pull on their star. This causes the star to wobble, which we can spot through clues in the starlight we see.

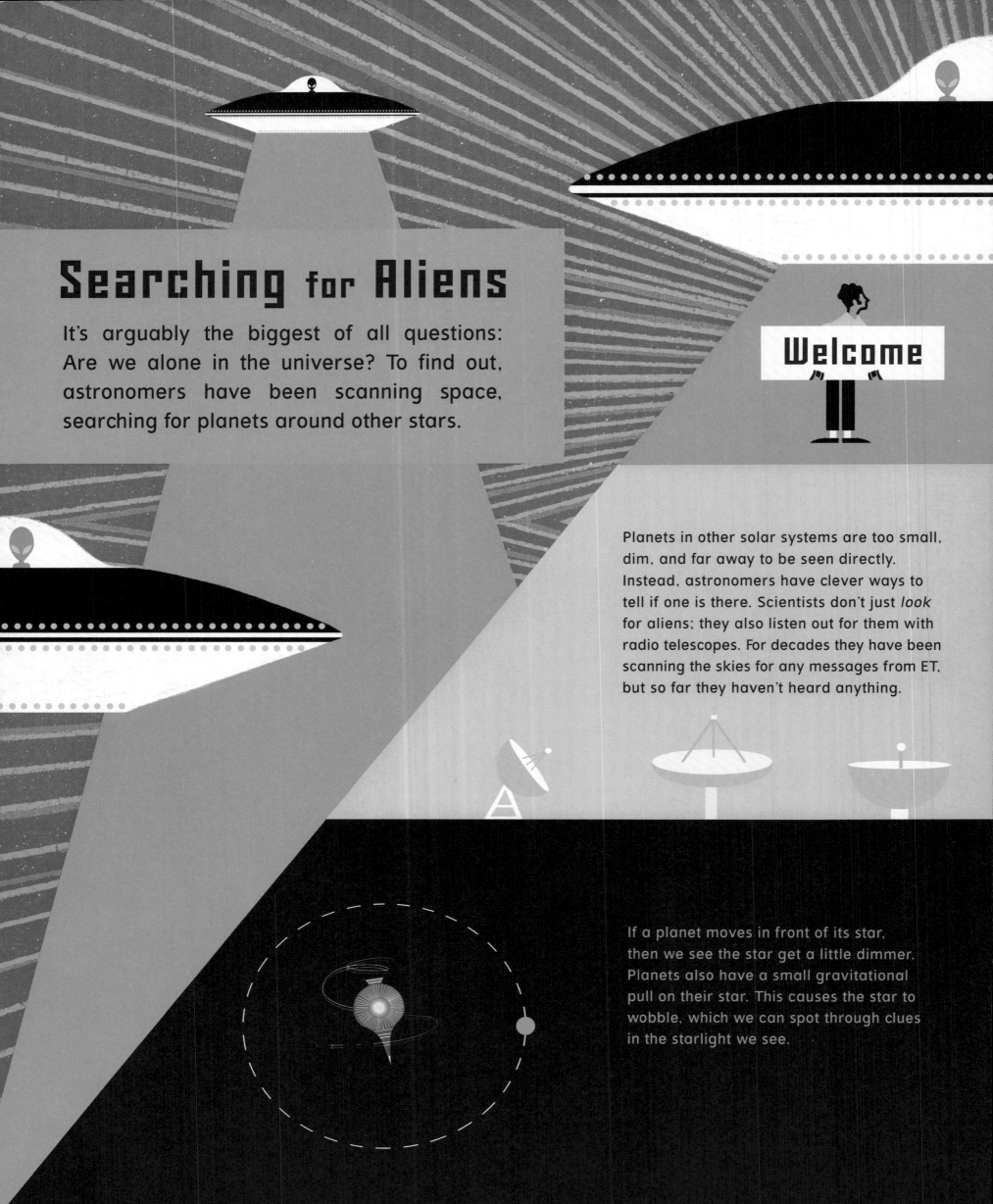

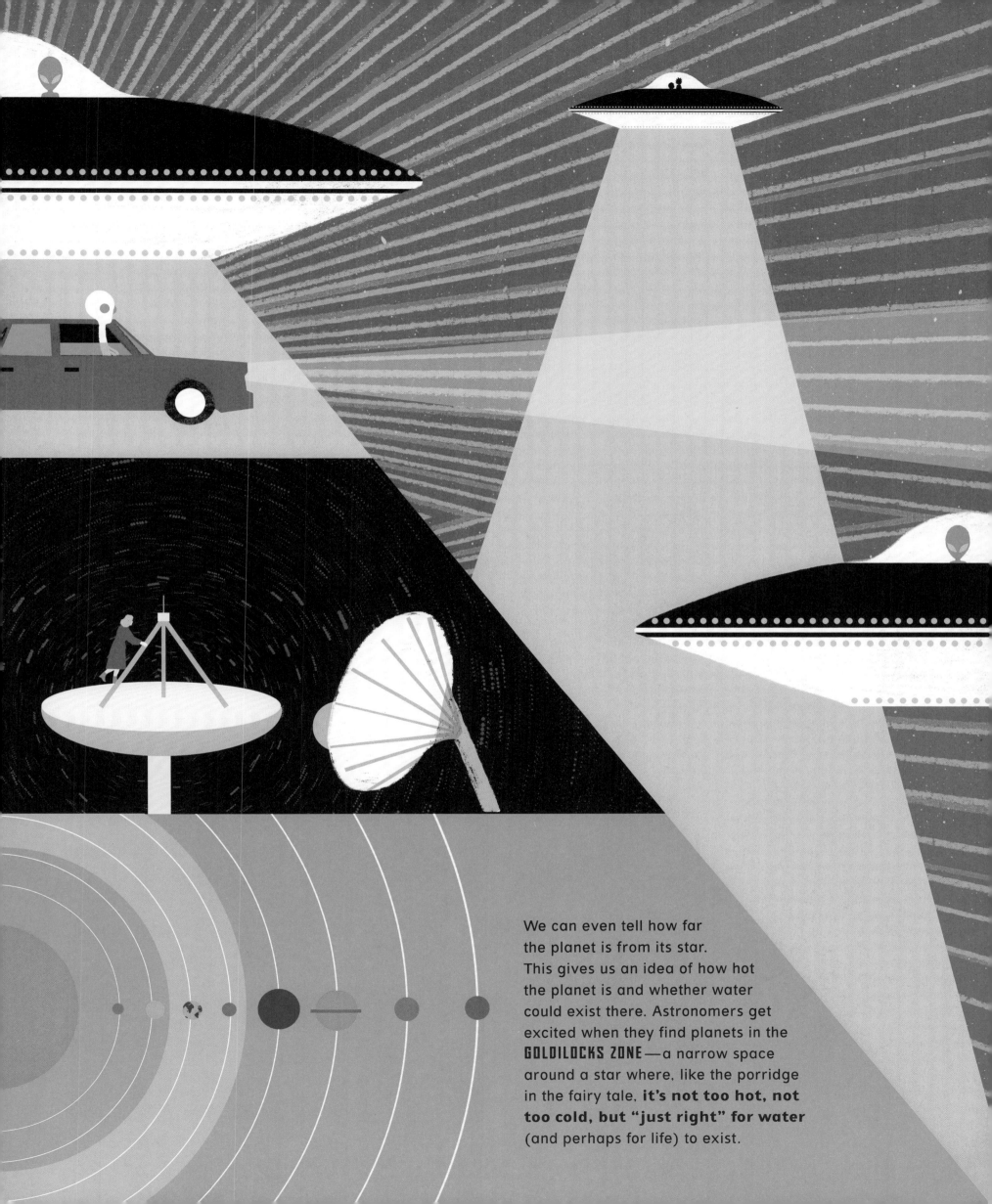

We can even tell how far the planet is from its star. This gives us an idea of how hot the planet is and whether water could exist there. Astronomers get excited when they find planets in the **GOLDILOCKS ZONE**—a narrow space around a star where, like the porridge in the fairy tale, **it's not too hot, not too cold, but "just right" for water** (and perhaps for life) to exist.

How Old Are Stars?

How do we even begin to work out the age of a star and what it's made of? After all, stars are so far away that we can't visit them. Even if we could, the temperatures would be too extreme to take a sample — even in a space suit, we couldn't get closer than 3 million miles/5 million kilometers to the sun before becoming toast. Astronomers have to think differently.

Stages of a Star

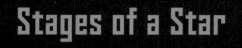

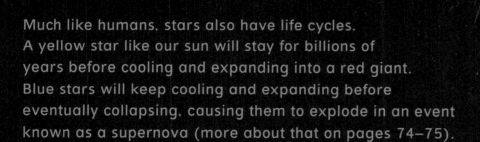

Much like humans, stars also have life cycles. A yellow star like our sun will stay for billions of years before cooling and expanding into a red giant. Blue stars will keep cooling and expanding before eventually collapsing, causing them to explode in an event known as a supernova (more about that on pages 74–75).

To measure the age of a star, astronomers have to think smart and use the only thing they have: starlight. If you pass starlight through an instrument called a **SPECTROMETER**—which is a bit like a prism—you can **split the light** into the familiar spectrum of rainbow colors. Look carefully, however, and you'll notice that some hues are missing. That's because different chemical elements in the star swallowed that color of light before it could head out into space. So this spectrum acts like a barcode, telling us exactly what a star is made of.

This "barcode" gives astronomers a neat way to age a star. In the early universe, the only ingredients around for making stars were hydrogen and helium. But as the universe has matured, more and more elements have been added to the pot. So a very old star is made of just hydrogen and helium. Younger stars have a much richer barcode, with more missing colors.

The Dark Universe

The universe is like an iceberg: the bit we can see is only a small part of what's really there. In the last few decades, astronomers have realized that they don't know what the rest is made of.

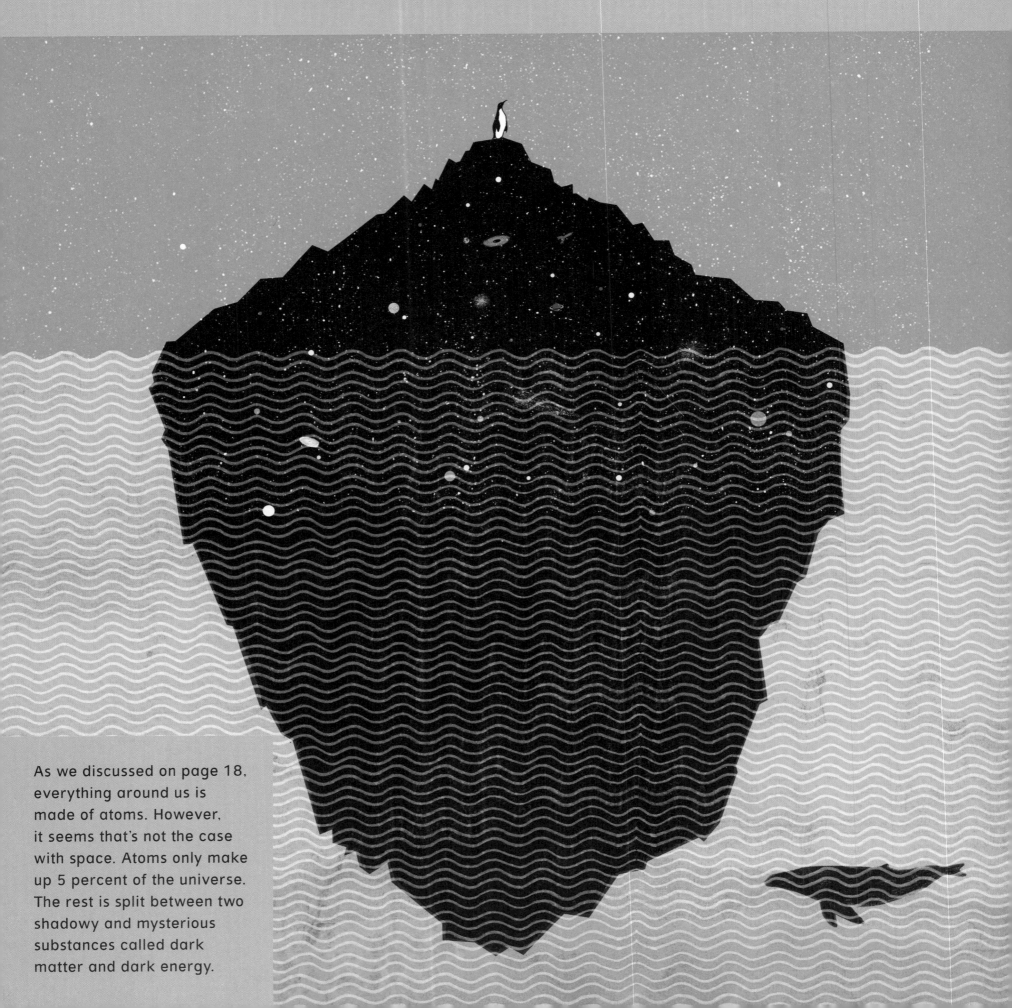

As we discussed on page 18, everything around us is made of atoms. However, it seems that's not the case with space. Atoms only make up 5 percent of the universe. The rest is split between two shadowy and mysterious substances called dark matter and dark energy.

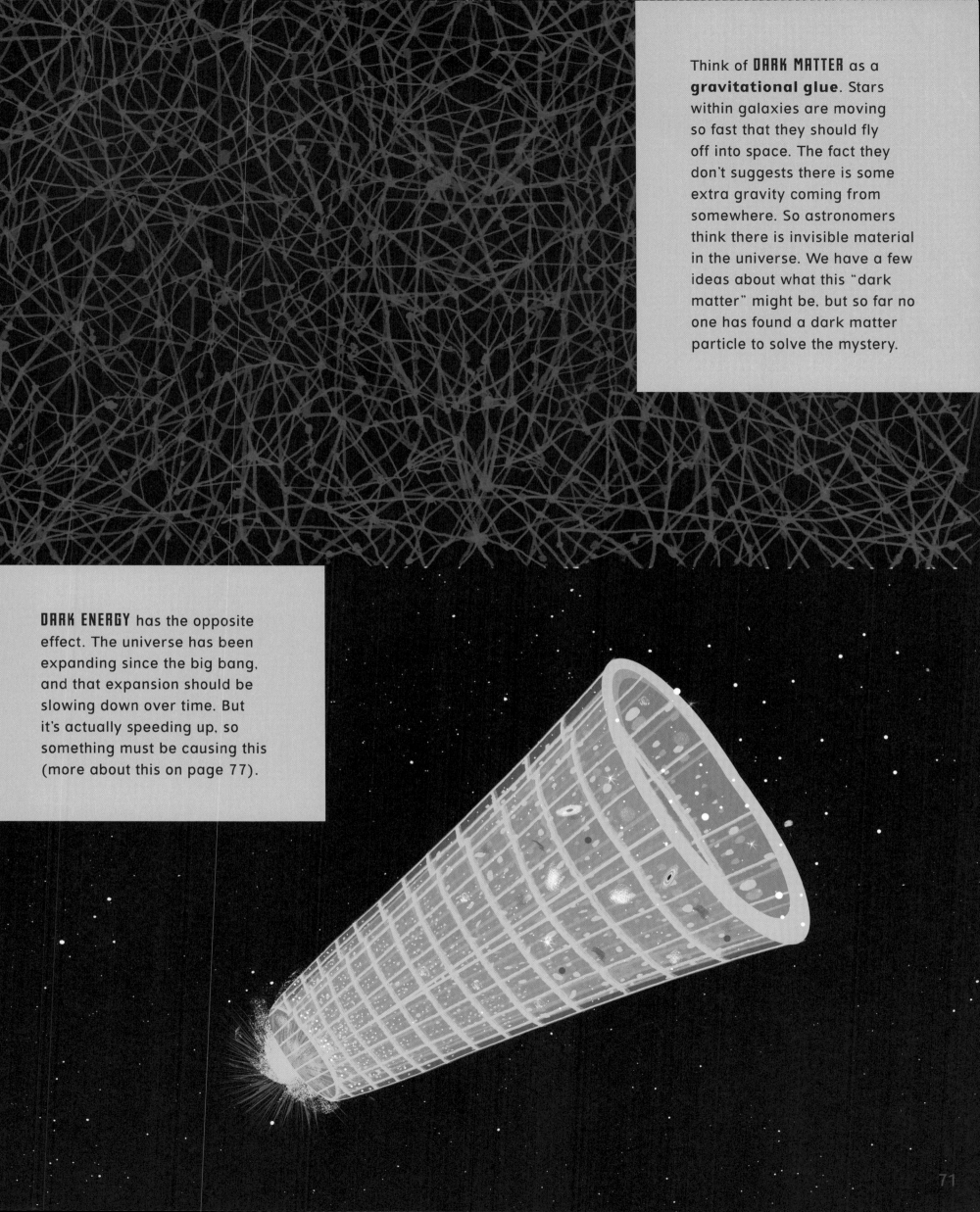

Think of **DARK MATTER** as a **gravitational glue**. Stars within galaxies are moving so fast that they should fly off into space. The fact they don't suggests there is some extra gravity coming from somewhere. So astronomers think there is invisible material in the universe. We have a few ideas about what this "dark matter" might be, but so far no one has found a dark matter particle to solve the mystery.

DARK ENERGY has the opposite effect. The universe has been expanding since the big bang, and that expansion should be slowing down over time. But it's actually speeding up, so something must be causing this (more about this on page 77).

Black Holes

You can only see these words because light is bouncing off the page and entering your eyes. Imagine instead that the book swallowed any light that hit it. You'd be left in the dark as to what's inside. That's exactly what's happening with a black hole.

When a really big star dies, it warps the space around it to such extremes that any light trying to escape ends up curving back in. To successfully get away from a black hole, you would need to travel faster than the speed of light, which we know isn't possible (see page 44).

So what would happen to you if you were unlucky enough to fall into a black hole? The answer is not good. The difference in gravity between your feet and your head would be greater than the bonds between your atoms—which means you'd **get stretched and pulled apart**. Physicists have a word for this process: **SPAGHETTIFICATION**. At the moment no one knows what would happen to your spaghettified atoms at the bottom of the black hole.

Anything could become a black hole if it were compact enough. Shrink the Earth down to the size of your thumbnail, and light wouldn't be able to escape. This is because it would become too dense for anything to break free of its gravitational field.

73

Standard Candles

When a star reaches the end of its life cycle and is extinguished, all that's left is a **small core** the size of Earth called a **WHITE DWARF**. The white dwarf becomes locked in a gravitational dance with its nearest star neighbor. It starts stealing gas and bulking itself up. But it gets too greedy, eats too much, and explodes with an unimaginable force and a searing light that can be seen halfway across the universe.

These cataclysmic events—called **TYPE IA SUPERNOVAS**—allow astronomers to measure the distances to faraway galaxies. **White dwarfs explode when they have roughly the same amount of material as 1.4 suns.** This is called the **CHANDRASEKHAR LIMIT**, after the nineteen-year-old Indian astrophysicist who calculated that number during a three-week boat journey to Europe in 1930.

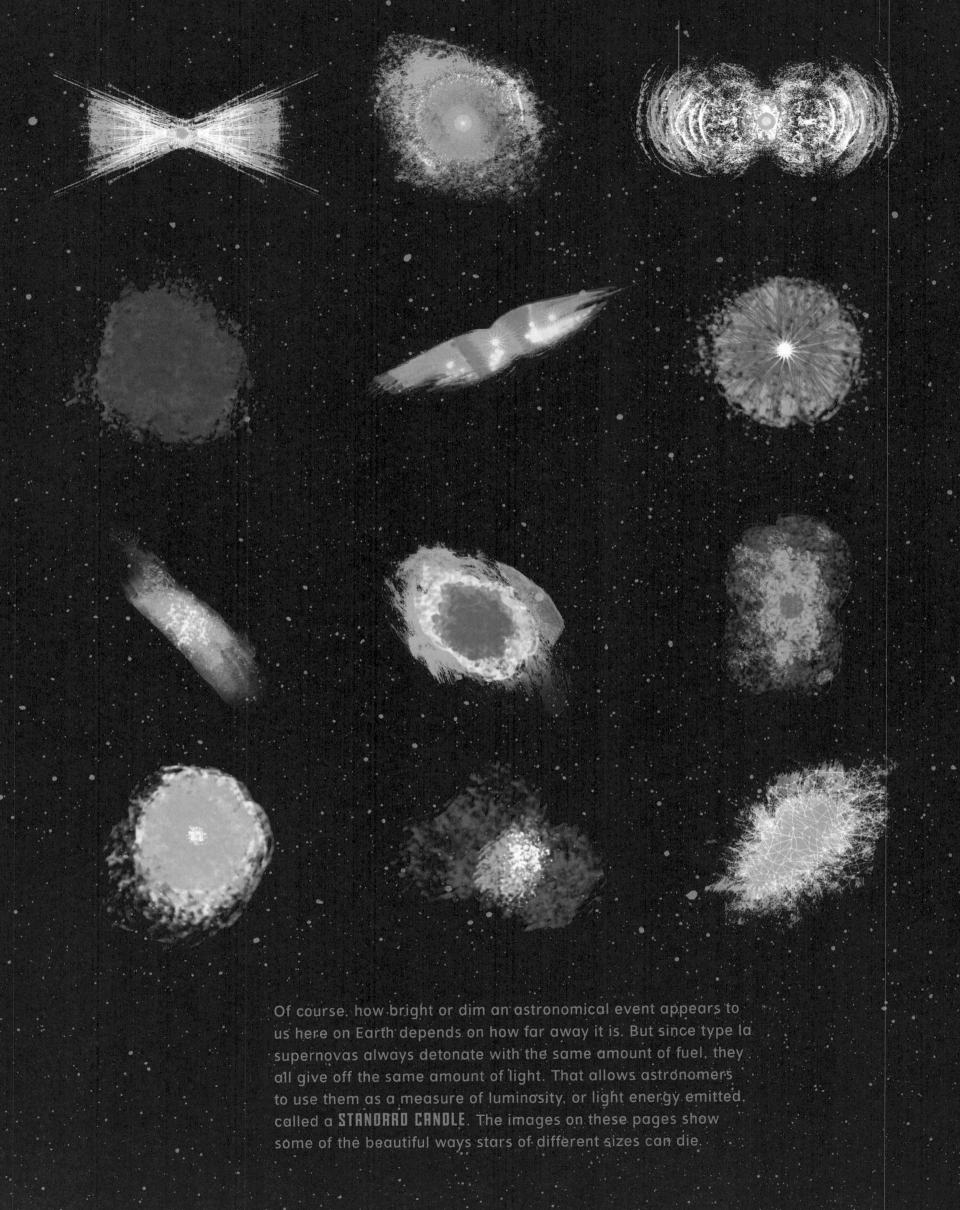

Of course, how bright or dim an astronomical event appears to us here on Earth depends on how far away it is. But since type Ia supernovas always detonate with the same amount of fuel, they all give off the same amount of light. That allows astronomers to use them as a measure of luminosity, or light energy emitted, called a **STANDARD CANDLE**. The images on these pages show some of the beautiful ways stars of different sizes can die.

The Expanding Universe

Have you ever heard an ambulance race by on the way to an emergency? Did you notice that **the sound of the siren changed** as it tore past you? This is called the **DOPPLER EFFECT**, and astronomers used the same idea to discover how our universe got started.

As an ambulance hurtles toward you, the sound waves it emits get squashed together—their wavelength gets shorter, and so the sound becomes higher pitched. Then as the ambulance rockets away from you, the sound waves begin to stretch out again. This shifts the sound of the siren back to a lower pitch.

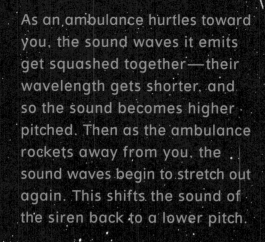

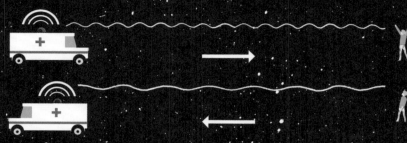

Light is a wave, too, so the same thing happens to moving light sources, except it is not the pitch that changes but the color. Light sources **approaching** us appear bluer, while those moving **away** from us appear redder.

In the 1920s, astronomers noticed that the light from almost every galaxy in the universe is "redshifted," meaning that it's redder than it would be if its source were at rest. This indicates that the other galaxies are moving away from us. Astronomers concluded that our universe must be expanding and that that expansion started with **a big bang nearly 14 billion years ago**. This is known as the BIG BANG THEORY. Dark energy is now making this expansion speed up, and it remains one of the universe's greatest mysteries.

Aside from dark energy, there is much more to be explored. We have yet to discover if life exists on other planets, or indeed what happens at the bottom of a black hole. Who knows what we might discover next and where it may lead us?